INVENTIVITY

THE ART AND SCIENCE OF RESEARCH MANAGEMENT

INVENTIVITY

THE ART AND SCIENCE OF RESEARCH MANAGEMENT

JOHN J. GILMAN
Berkeley, California

VAN NOSTRAND REINHOLD
New York

Library of Congress Catalog Card Number 92-3239
ISBN 0-422-01186-5

Manufactured in the United States of America

Published by Van Nostrand Reinhold
115 Fifth Avenue
New York, New York 10003

Chapman and Hall
2-6 Boundary Row
London, SE1 8HN, England

Thomas Nelson Australia
102 Dodds Street
South Melbourne 3205
Victoria, Australia

Nelson Canada
1120 Birchmount Road
Scarborough, Ontario MIK 5G4, Canada

16 15 14 13 12 11 10 9 8 7 6 5 4 3 2 1

Library of Congress Cataloging-in-Publication Data

Gilman, John J. (Joseph)
Inventivity: The art and science of research management / John J. Gilman.
p. cm.
Includes bibliographical references and index.
ISBN 0-442-01186-5
1. Research—Methodology. 2. Inventions. 3. Creative ability in technology. I. Title.
Q180.55.M4G42 1992 92-3239
658.5'7—dc20 CIP

To

Gretchen Sutter Gilman

There is nothing more difficult to take in hand, more perilous to conduct, or more uncertain in its success than to take the lead in the introduction of a new order of things, because the innovator has for enemies all those who have done well under the old condition, and lukewarm defenders in those who may do well under the new.

—N. Machiavelli, Il Principe (1513)

Contents

APPENDICES

PREFACE

You might well ask, "What is inventivity?" It would not be surprising if you didn't know the meaning of this word. I didn't know either until I coined it in 1979.[1] The word is a sister of "productivity." It means the efficiency with which inventions are made. That is, the number of man-hours, or dollars, needed to produce an average invention. This is a far more important parameter than an aggregated number of inventions, or a number of patents per capita of the general population.

Although there are inventions that are not represented by patents, the only objective evidence that a particular individual has made an invention is the issuance of a patent. Thus patents are a measure of proven inventions, which may, or may not, have been reduced to use in commercial operations.

The word "invention" is often used in a broader context than patent. Invention may refer to conceptual theories, ideas in the decorative arts, passages in literature, novel computing programs, and so on. Such inventions have costs associated with them, so the notion of inventivity might be applicable, but the viewpoint of this book is limited to the practical arts. That is, to novel advances that could be patented, in principle. Whether a particular invention actually becomes patented depends on various factors, including its relationship to previous inventions, the cost of developing it, the potential market for it, and others.

I became interested in inventivity several years ago when I was given a list of the number of patents produced in 1977 by various companies. This list was given to me by the Patent Department of the Allied Chemical Corporation (now Allied-Signal). As would be expected, the list showed that the largest industrial companies produced the most patents. But I noticed that small companies seemed to produce more patents than would be expected from simple proportionality. So I plotted the data according to various types of distribution and found that, for a given amount of sales, small companies produced more patents than large ones. This surprised me because I had thought that the most effective producers of inventions were the large chemical, electrical, and telecommunications firms. One might conclude that I was a victim of advertising.

I was aware that small entrepreneurial firms are often more successful than large ones in bringing new products to the marketplace, but I had not connected this with their relatively greater efficiency in making inventions. Furthermore, I was surprised at the strength of the inverse correlation. In some early plots there was direct inverse proportionality between the number of patents issued to a firm in a given year divided by its sales (that is, its inventivity parameter), and its sales! Later, for more critically collected data, the power law coefficient of the correlation became less than one, but a strong inverse correlation remained.[2]

Since the fraction of their sales that companies spend on research is very roughly constant, this correlation indicates that the cost of producing a patent (the inventivity) varies markedly from one company to another. Why? It seems unlikely that size alone is the independent variable. There must be other factors, related to or affected by size, that are important. These management factors are the subject of this book. In other words, this book is a discussion of research management from the viewpoint of maximizing the number of important inventions that are made per unit of cost. This, after all, is the mark of good management.

The total resources available to any society are limited, so the societies that lead in the critically important activity of making inventions will be the ones that have the highest inventivity. Thus it is the most important single measure of the effectiveness of research in determining the destiny of a society. There are other measures that some members of the research community consider to be very important (prestigious prizes, high-ranking jobs, and the like), but these are not so important to the society at large. Nor, in the long run, are they so important to the research community itself.

Although great discoveries are sometimes made with limited resources, their effect on the society that directly surrounds them is limited unless multitudes of derivative inventions are made. There are many examples of discoveries about which it is said that they were "made before their time."

Many organizations are not satisfied with the productivities of the research groups that are associated with them. These groups may be either internal departments or external people with whom work is contracted. The dissatisfied organization may be a manufacturing company, a university, or a government agency.

The typical response of the dissatisfied organization is to try to adapt knowledge of the factors that affect manufacturing productivity to the research group. After all, productivity is just output rate divided by input rate isn't it? The input rate is well-known in terms of dollars, so all that remains is to optimize the output rate. The usual result of this approach is alienation of the research people followed by decreasing morale, leading to a drop in real productivity. Why doesn't it lead to the desired increase in productivity (inventivity)?

The answer is associated with the nature of the "output." This is profoundly different for creative organizations than for ones that produce hardware or standard services. The difference is not a quantitative one; it is qualitative. To clarify this, compare how the outputs from a simple manufacturing operation are distributed with how outputs are distributed from a research laboratory.

Suppose the manufacturing operation makes pins of a certain standard diameter. The output will be distributed normally about a mean nominal size. Some of the pins will be smaller, and some larger, than the nominal. The better the operation, the smaller the deviations from the nominal. Quality will be controlled by accepting only those pins whose diameters lie within an upper and a lower limit. The relatively small number whose diameters lie outside the limits will be rejected.

Now consider the research organization. Its output will be distributed quite differently. Its purpose is to produce novel concepts or new arrangements of old concepts, leading to new devices, materials, or processes. Thus its output must have a skewed distribution because old concepts are excluded from it, and because it must cutoff when "novelty" becomes too small to be significant. Novel concepts do not all have the same importance so there is a distribution of them about some mean level of novelty. But in this case quality is not determined by how close to the mean a new concept is. Instead, it depends on how far from the mean it is! Attention is no longer focused near the mean of the distribution; it is focused at the extremum. Concepts near the mean are now rejects, while those that lie far from it are acceptances. This is indeed profoundly different from the manufacturing case. The two cases are qualitative opposites.

The difference between the output distributions as outlined causes many of the difficulties of research management. This difference also leads to the communication problems that arise between research organizations and the senior managements that oversee them.

Perhaps the astounding thing is that anyone would even try to apply knowledge of one of these situations to the other. But it is done regularly. When an attempt fails, it is said that there must be a communication problem. Nonsense. Inept treatment of a problem has little to do with communication in the usual sense. Something deeper is involved.

But, if principles of manufacturing productivity do not relate to research productivity, then what principles do? That is the subject of this book. No doubt the discussion given here will be incomplete; however, it should serve as an introduction and will attempt to survey the best available thoughts. The reasoning will be largely statistical because that is the only means for approaching such a complex subject quantitatively. To approach it anecdotally would be to follow past discussions and would not advance the subject toward a discipline.

To some extent there is an inconsistency in taking a statistical view of invention because important ones are rare and sometimes unique. But this book is about the management of invention, not about the inventive process itself. In other words, it is not about the creative process, but about improving an organizational environment in order to encourage creativity.

Communication was previously mentioned. When a research organization is immersed within a larger organization, industrial or otherwise, it needs to communicate with its environment. Such communication is often recognized as difficult, and the blame is commonly laid at the feet of the researchers. This puts them on the defensive; a prudent response in the face of greater political power. However, it also inhibits their daring, so the overall organization loses, then wonders why its research activities are ineffective!

The cause of the difficulties in communication between the research and the general management is the marked difference in the values of the types of work they do and the distinctly different motivations for doing it. General management considers itself to be a team in competition with other teams to conquer a territory. There is much talk about "making money for the stockholders" of the company, and providing societal services, but little evidence that these are the underlying motivations. When an analogy is needed, general managers usually start talking about football or a military campaign. For them the competition is clearly defined. It is another company, or a rate-setting board, or a government regulatory agency, or some other invader of their territory.

By contrast, first-line researchers have little motivation to conquer an old territory. They want to open up new territories. And they want to do it themselves. They want to create an embryo that will grow to become a territory. This is solitary work. Even if it is done with a large-scale machine like a particle accelerator, it is done by a solitary "corporation."

The success of a general management team is usually attributed to the company that employs the team; in the research arena it is associated with an individual person. Development work is another matter because it is necessarily a team effort. Researchers are concerned with learning, conceiving, discovering, and understanding. And with recognition of their skills in these areas. Thus their reward menus are quite different from those of general managers.

Given these differences, it is easy to see why these two disciplines do not communicate with one another readily, except at a superficial level. It should also be clear that neither group can get good communication from the other by arbitrarily demanding it. In fact, it is probably true that, if it is perceived that they do communicate well within a given organization, then the functions that they should be performing are not being done as effectively as they should be.

Various manifestations of the communications problem are discussed in the chapters that follow. This book will, I hope, help improve communications through improved insight. I also hope that some of the factors that determine inventivity will be illuminated. This parameter can be increased either by increasing output for a given amount of input resources, or by reducing the input needed for a given output.

NOTES

1. For example, see J.J. Gilman, Physics Today (October 1980) p.9.
2. J.J. Gilman and A. Siczek, "Optimization of Inventivity," Research Management (July–August 1985) p.29.

ACKNOWLEDGMENTS

In addition to his direct contributions in the form of the conversations reported in the text, George Kazan contributed immeasurably to this book through lengthy discussions of the issues, critical readings of manuscript pages, and encouragement. Gretchen Gilman helped in similar ways. George Duvall made incisive comments that added a different perspective. Mario Rabinowitz provided cogent points regarding Pareto distributions, as did Albert Holt regarding the market share rule. The quotation used for the frontpiece was brought to the author's attention by Carl Cline, and confirmed by Richard Robinson. Publishers' reviewers made useful criticisms, both constructive and otherwise. Renee Ford and Arthur Miller provided encouragement and facilitation when it was most needed.

1

NATURE OF A RESEARCH ORGANIZATION

Imagination is more important than knowledge. For knowledge is limited, whereas imagination embraces the entire world, stimulating progress, giving birth to evolution.

—Albert Einstein

Most of the literature on research management is anecdotal. It describes a great variety of case histories—mostly in the success category. This generates an impression that research management is an art with few, if any, general principles or techniques. To a large extent this impression is correct; however, it is incomplete because there are some rules and methods that can be applied quite beneficially. They do not seem to be well-known so one of the purposes of this book is to collect and describe them in the expectation that they might find more extensive appreciation and use.

One of the reasons it is difficult to find general management principles that apply to research organizations is that they are so difficult to manage. I am using the term manage in the sense of organizing and leading a group of people with a common purpose. I do not wish to emphasize the administrative part of managing. Although this is an important part of the total process, it is not the central or essential part.

Research management is also intrinsically different from production, marketing, or finance management. The purpose of research is to generate new knowledge, and to make inventions as well as innovations. Therefore, its essential activities interface with the unknown and the future—not with the past, or the present, or the known. Many day-to-day research activities are the same as those of other functions, but these are not the essential activities. To paraphrase Thomas Edison, it is inspiration that is essential, not perspiration, even though inspiration occurs rarely.

To gain improved understanding of complex situations, it is often useful to consider simpler models. In the case of research organizations, I think that musical ensembles form good analogies. Especially jazz ensembles. Why? Because their purpose is to create new music in real time. Classical music ensembles recreate music that a composer has created at some previous time.

The best jazz musicians invent new musical phrases in real time with amazing proficiency. They also discover new sounds by pushing at the frontiers of their instruments. They push higher, lower, faster, smoother, rougher, with greater rhythmic complexity, and so on. These are similar to the things that research workers do, except that for them the results are new materials, processes, devices, systems, theoretical relationships, and so on.

Readers who have observed, or perhaps played in, jazz orchestras know that the people in them are very unlikely candidates for jobs as production managers or manufacturing workers. Although jazz is played according to well-defined harmonic and rhythmic rules, and although the skills of the musicians are often exquisitely meshed, a given composition is never played in exactly the same way twice. Thus a jazz ensemble does not "produce" music. It creates it almost continuously.

The principal purpose of research is learning. This may consist of new abstractions as in the creation of new theories, or it may consist of concrete inventions such as new devices. As research is pursued, old and new pieces of knowledge are brought together in the mind of the researcher and combined. Out of this synthesis comes new learning. Since the nature of the learning is not known in advance, the process is open ended, and therefore is an art form. Being an art form, the invention process creates social troubles for research organizations because art is the antithesis of the production of goods and services. But most of the people in a society are engaged in doing the latter. This is most clear in the case of industrial firms organized specifically to produce uniform products. An ideal manufacturing firm acts like a programmed computer, printing out endless rows of codified digits. Contrasting this, an organization designed to practice and develop an art form tries to make each product as distinctly different from its predecessors as possible.

The elements of an ensemble of people that are needed for the creation of either new music or research results are as follows:

1. Leading performers (inventors)
2. Supporting performers (experts)

3. Arrangement (organization)
4. Conductor (coordination)
5. Instruments (apparatus)
6. Performance hall (laboratory)

All of these must be first rate, not just a few of them. Otherwise, the results will be mediocre.

The most important single element of this list is the leading performers. If the players of the lead instruments in a jazz ensemble are not exceptionally talented, no music of consequence will be created, regardless of the qualities of the other elements on the list. Creative performers are characterized by intense motivation. This has usually been awakened in them during an apprenticeship under a past master of their instrument. Creative performers create new musical or technical ideas by assimilating existing ideas and combining them in new ways. Thus they need to have large capacity memories that are somewhat "leaky," or disorganized, so that ideas do not become too firmly entrenched in them. These somewhat leaky memories allow items from some category to be brought into contact with items from other categories during the creative process. The minds of such people are quite different from the legalistic, tightly organized minds that are best for preserving and operating with preexisting knowledge. Both types are necessary, however, for the long-term health of an organization.

Inventors of music or technology must also have high tolerances to risks, and to be able to accept imperfections. The most creative musicians push their instruments beyond the state of the art. This sometimes results in wrong notes, poor phrases, or other failures. Similarly, inventors and their managements must expect and accept many failures to precede most successes. If they do not occur, the organization is not pushing the state of the art hard enough.

Leading performers alone are not enough to make either music or inventions. I witnessed a forceful demonstration of this some time ago at a Charles Mingus and Friends concert. Mingus was a leading bass viol player and jazz composer. One of his friends was Bill Cosby, the comedian. At the beginning of the concert, Cosby came onto the stage dressed as a funky conductor and announced that he would conduct his rendition of the jazz classic, "Honeysuckle Rose." Then each instrumentalist began to play the melody. Each in his own key and with his own rhythm. The result was cacophony. Thus, a creative ensemble needs the guidance of a musical arrangement, or an organizational structure, that has a beginning, a middle, and a purposeful result in sight.

Another important point is that the efforts of lead performers can be greatly amplified by strong supporting performers. This is clearly illustrated in the world of music by the rhythm instruments that support the lead voices. A poor drummer, or bass viol player, can ruin an otherwise promising ensemble. Conversely, a superior rhythm section can convert an otherwise good ensemble into an outstanding one. I recall, for example, attending a performance at which the music of the outstanding pianist, Jelly Roll Morton, was being recreated by a fine musician, Bob Greene, together with a supporting group. The entire group, and especially Greene's piano playing was excellent, but the thing that really brought joy into the performance was the driving, vivacious playing of the bassist, Milt Hinton. Similarly, first-rate technical, analytic, library, and shop services can mean the difference between a good and an excellent research organization.

Many pressures act on and within any organization. These will tend to randomize its efforts if they are not counteracted in some way. This is the job of the conductor or director. If the organization functioned ideally a conductor would not be necessary, but in reality the director must coordinate its various activities, and continually restrain wayward trends. By this means the overall effort is kept moving toward a desired goal, or toward a set of goals. Also, the conductor, or director, ensures the research organization's continuity by recruiting and hiring new members to replace old ones, or expanding the number of people in the organization.

The final essential element in the operation of a creative ensemble is its facilities; that is, its housing and instruments. The housing cannot be treated casually because it strongly affects the interactions between the people within the ensemble and between the ensemble and its audience. For a musical group the acoustics and the general ambiance of a performance hall affect both the musicians and the audience. For a research group there are analogous effects on communications and attitudes. This includes effects on communications into the group, within the group, and outward to patrons of the group.

One element that has been left out of my list is central planning. Creative ensembles should have overall goals but not plans for detailed activities because these are ineffectual. General principles must be observed and honored, and some irreducible rules must be followed, but specific actions should be based on the seizure of new ideas or opportunities as they arise. Since both the results being sought, and the best plans for pursuing them are not known at the outset, it is not possible to plan them. An unpredictable process such as this severely frustrates people who like to follow predetermined plans, usually provided by some higher authority. However, the process cannot be avoided when the purpose of the work is to create something new.

Management consists largely of making selections. Good management selects good people, good technology, good ambience, and so on. Some people in management positions express the opinion that their function is to "make decisions." That is, they see themselves as judges, rather than selectors. In my experience, these people are bad research managers. Too often judgment involves eliminating the poorest options. Selection, on the other hand, involves picking the best options; often with inadequate information. In other words the judge decides what is wrong, if anything, but the selector has the far more difficult task of deciding what is right. Some persons have a talent for sensing at least a part of what the future holds, but it is a rare talent.

With an output that cannot be described initially because it does not exist, how can rational purposes, strategies, and tactics be explained? Often they can only be described in statistical terms. This is not the case for development and problem-solving work, of course. Although it is conventional to couple research with development through the acronym, R&D, this is like coupling salt and pepper rather than sugar and cream.

Because of the operation of the "rule-of-tens" (to be discussed later), about 90 percent of an organization's activities produce results that are not useful. How can anyone be convinced that the organization is adequately productive? The answer lies in the very large amplification that is provided by exponential growth.

It is difficult to transmit whatever universal content the anecdotal literature of research management has from one person to another. People who are inexperienced students of the subject will wonder how specific anecdotes should be interpreted because they tend to apply to the special circumstances of specific cases. Experienced people will be able to relate similar anecdotes themselves so they will not learn much that is universal from listening to another's anecdote. On the other hand, if the results of experience can be expressed in quantitative terms, their content becomes more readily understood. This is because parameters that are expressed as numbers have generality associated with them; otherwise they could not be numerically expressed. Although quantification can only be accomplished for a few topics at present, these are a start.

Doing research is a lonely pursuit. The best researchers like it that way. The less than best don't like it, and they often move to other activities. If they move into management, a conflict is born. On one side there are people who don't mind being on their own outside the mainstream. This is synonymous with being willing to take risks. On the other side are those who are risk averse with a strong need to be part of the central stream.

Taking risks means being unable to predict outcomes reliably. The higher the risk, the lower the predictability of the outcome. This is anathema to most business managers—for good reasons. However, when an outcome is realized that was not initially predictable, a unique situation arises that may have exceptional business or social potential. Searching for such opportunities is the highest purpose of both public and private research.

People who rise through the ranks of corporate and government management are those who are considered to perform their successive jobs in a uniformly "successful" manner. They can only do this by avoiding risks of all kinds, including those involved with interpersonal relationships, projects, and investments. It would be surprising if these people related well to risk-taking researchers, and mostly they do not. The exceptions are rare.

It is difficult for risk-averse managers to place value on people in proportion to how little they fit into a society instead of how well they fit. For a consensus style of management this creates a virtually insoluble problem.

As Chomsky[1] and others have pointed out, one of the reasons why particular managers rise through a hierarchy is that they accept the precepts of the organization. They are "company men"; not necessarily because they connive to be so, but because they have bought into the styles and values of the organization, as well as those of its external social environment. This happens perhaps because they have accepting temperaments, or perhaps because their early education has taught them to do so. They may have a strong need for acceptance by the organization, so there is a reciprocal relationship between them and it.

The inventor is different. This is suggested by G. B. Shaw's quotation at the beginning of Chapter 4. He examines precepts independently of the organization in which he resides and decides for himself whether they seem valid in the context of the present conditions. He is a skeptic.

In order for the manager and the skeptic to communicate, their differing attitudes must be known and recognized. The burden for this is the manager's, but it is rarely accepted by him. The more usual approach by far is for the manager to attempt to subordinate the values of the maverick. This is one reason why large research organizations tend to be dull; unless their component parts are decentralized.

NOTES

1. Naom Chomsky, *Bill Moyers—A World of Ideas*, ed. B. S. Flowers New York; Doubleday, (1989), p.51.

2

WHY IS RESEARCH WORTH DOING?

Businessmen go down with their businesses because they like the old way so well they cannot bring themselves to change.... Seldom does the cobbler take up with a new fangled way of soling shoes, and seldom does the artisan willingly take up with new methods in his trade.

—Henry Ford *My Life and Work*, 1922.

The creation of new assets is the broad purpose for doing most research. The new assets range in character from intangible intellectual systems to highly tangible consumer products, tools, or software. They may be used to organize knowledge, be consumed, as a basis for developing still newer products, for making processes more efficient, for improving the productivity of people, for trading with other companies, for doing further more sophisticated research, and so on. They may be hard assets such as buildings, bridges, minerals, and manufacturing machinery; they may be soft assets such as organizations, skilled people, and money. They may be stored information: books, tapes, disks, films, bacterium strains, DNA sequences, and the like. Or, they may be stored knowledge: patents, science, skills, and so on.

Some people think that research should be done principally to satisfy curiosity, or to buttress a theological position, or to advance culture generally.[1] Indeed some of it should be done for these purposes. But a more legitimate reason for doing most research is to make inventions that are of interest to society and tend to advance civilization.

Doing research may have begun with finding ways to make fire and tools for hunting.[2] There was little of what is now known as science. Flint stones were being skillfully shaped through chipping long before the mechanics of fracture became a science. As toolmaking progressed, it was discovered that research could lead to new operational methods such as doing calculations,[3] drafting, timekeeping, printing, navigation, and other activities important to commerce and government (the importance of information is nothing new). It was found that methods for making things could also be improved, as well as the military arts. New materials, energy sources, and communication systems were created. In a variety of ways research work led to asset generation, and merchants soon realized that one way to increase their assets was through the use, that is the development, of research results.

In order to survive, merchants must have assets at the end of a year that equal, or preferably exceed, those they had at the beginning of the year. In order for their business to grow significantly, their assets must in fact increase substantially from year to year. Asset generation thus became one important purpose for doing research within businesses.

Services are also often enhanced by the results of research; an outstanding example being health care. In this case knowledge and skills are the principal assets. These are "soft" assets and are more difficult to evaluate than the "hard" assets that are so important in manufacturing or merchandising. But they are no less important.

It is sometimes said that the purpose of research is to create change. It does do this, but change for the sake of change is of doubtful value. This is why I have placed an emphasis on asset creation.

Research may also produce military entities. These are a special kind of asset. If they act as a deterrent to war they have indirect positive economic value. On the other hand, in the event of war, they are used by both sides to destroy the assets of each other. Thus, in this case, they have negative economic value. To the extent that their costs exceed the amount needed for deterrence, they are a sunk cost with zero return.

One of the best measures of the performance of a business, or of a society, is the return on assets (that is, the rate of increase of assets per unit of asset). In the absence of debt, if the return is kept internally, this equals the growth rate.[4] Few societies want their growth rates to be other than positive so they strive to increase their assets each year. There are various means for doing this, including: acquiring assets with high rates of return; cutting costs while holding the asset base constant; and creating them through research and development.

Unfortunately, the word research does not have a precise meaning. As a result, before the topic of this chapter can be discussed clearly, it will be necessary to define it for the present context. According to *Webster's Ninth New Collegiate Dictionary* it means "careful or diligent search." This is a good definition, but it does not tell us how this differs from "exploration," or from "investigation," or even from "study." Also, how does it differ from "search" as in looking for a lost object. And this definition appears to exclude novel experimentation.

A somewhat different, and I think better, definition is given in the *Random House Collegiate Dictionary (1980)*, namely; "systematic inquiry into a subject in order to discover or revise facts, theories, etc." This gives good coverage of academic research, but does not cover industrial research very well. Perhaps this illuminates a problem in the communications between workers in these two major research communities. The problem is that although the name is the same, the content of the research activity is different. The difference in content is centered on the fact that different social values reside in academia compared with industria.

Since the values of academic research and industrial research are markedly different, it should not be surprising that the two communities have trouble communicating beyond superficial technical conversations and discussions of the future employment of students. Academic research is concerned with relating various phenomena to one another in order to form patterns. This requires *analysis* and is closely related to the teaching functions of academic institutions. Without relational patterns of phenomena to condense myriad facts, it would be virtually impossible to pass comprehensive knowledge from one generation to the next. Each generation would almost need to start over in the assimilation of large numbers of facts. Relational patterns (theories) are also very useful for design work, data management, guidance of future research, etc. Finding relational patterns is not the only concern of academic research, but it is the principal concern. Neither is finding relational patterns purely a matter of analysis. Sometimes new parameters, or techniques, are created. The parameter we call energy is one example; the technique of differentiation is another. Nevertheless, the central activity is analysis.

In contrast, the primary concern of industrial research is invention; that is, *synthesis* to form new techniques, materials, devices, or systems. Not all inventions are of interest; the only ones that qualify are those that have potential for substantial economic utility. Sometimes a synthesis, or invention, can be accomplished through pure thought. More often it involves systematic, and intelligent, experimental studies. Also, the importance of accidental discoveries should not be underrated.

It appears that tangible asset production was the goal of the earliest researchers. They sought convenient shelter and better tools for acquiring food. No doubt they observed some correlations between the variables of their environments and took advantage of these whenever it was possible, but activities directly associated with survival must have had the highest priorities. Along with the development of tools for domestic purposes, the development of weapons occurred. The function of weapons soon became ambiguous. They were used against animals for obtaining food, but also against humans for conquest.

Gradually, observational and recording tools appeared. These allowed more general, as well as more complex, correlations between the variables of the world to be found. The value of these tools for planning and designing was soon recognized and searches for additional correlations began to be pursued systematically. Surveying problems of various kinds led to the systematic study of geometry. This yielded concepts like the squares and square roots of variables. Correlations were then found between these derivative variables that allowed solutions of problems to be engineered, and thereby accomplished more efficiently than had been previously possible.

This early interplay between societal needs for tools and products (technology) and systematized knowledge (science) has continued with increasing intensity to the present. And it shows no signs of abatement overall; although the intensity waxes and wanes within specific parts of the total.

Technology tends to be the leading activity as it generates tools and large arrays of observations, which become correlated and systematized later. This is virtually by definition because observations cannot be correlated and systematized prior to their existence. Also, technology generates a majority of the advances in the tools of observation and record keeping that are needed before advances can be made. Technology does not appear spontaneously, of course. Its advances depend on previous consolidations of knowledge. Therefore the interplay of the two is essential to the health of both.

Finding simple cause-and-effect relations between doing research and realized benefits is extraordinarily difficult because of the large amount of time that often elapses between doing the research and the time when benefits may be recognized. Also, it must be noted that much research does *not* lead to recognizable benefits. Nevertheless, in addition to subjective observations, there is statistical information that connects initial research with eventual benefits. This will be discussed after some less abstract information has been mentioned.

ELECTRICAL INDUSTRY

The first sizable, organized effort to do industrially oriented research was that of Thomas Edison.[5] Others have claimed the distinction,[6] but their cases are weak. The Edison organization was extraordinarily successful. It not only invented new technologies but also established new businesses. Among its successes were incandescent electric lamps, central electricity generation and distribution systems, practical telephone transducers, motion picture systems, and phonographs. The assets and incomes generated by these developments dwarfed the original research costs. More will be said about this in a later chapter.

SYNTHETIC POLYMER INDUSTRY

While the Edison organization invented electrical devices and their connective systems, the chemists who worked in research at the E. I. du Pont de Nemours Company invented polymeric chemicals from which superior fibers, films, and moldings could be made.[7] The contributions that these inventions made to the earnings of the company are shown in Figure 2.1 (note that a few of the inventions such as tetraethyl lead, freons, etcetera were not polymeric). It may be seen that something new was commercialized about every five years with rest periods in between. Nylon fibers alone contributed an average of 166 million dollars per year to earnings over the given ten-year period. This was nucleated by the efforts of a few chemists under the direction of Wallace Carothers in the early 1930s.

Although the nucleating research events are only a small part of the overall scenario associated with a new commercial product, their considerable power is well illustrated by the chart of Figure 2.1. Even when they were in decline, products such as auto finishes and cellophane were still contributing to the earnings of the DuPont company 40 years after their initial commercialization. Saturated selling prices in the face of rising raw materials costs made margins thin for the older products, but these are operating problems, not coupled with the original R&D work.

PHARMACEUTICALS

The benefit that is most directly measurable is monetary return on a business investment. However, the correlation between return and research investments is not simple because several other factors affect the return. These factors act between the time that the investment is made and the time that

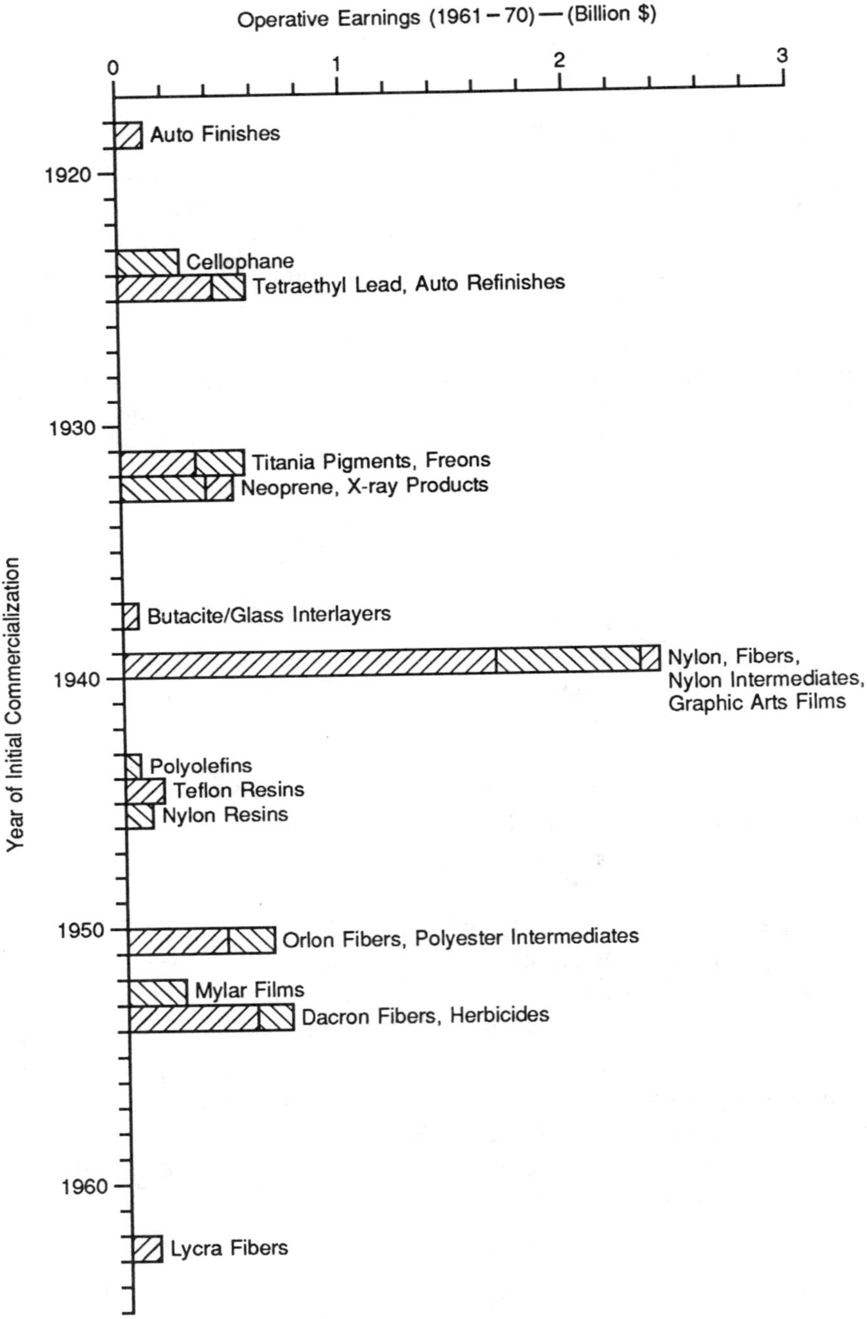

FIGURE 2.1 Major product groups of the DuPont Company 1961–70. Their cumulative contributions to operative earnings (before overhead and tax deductions) are shown together with the years in which they were first commercialized. (Source: Hounshell and Smith, "Science and Corporate Strategy").

the return is realized. This period may be 15 years or more. Return on investment is relatively short in the case of pharmaceuticals. This makes it possible to correlate the new sales of pharmaceuticals that are generated by earlier research expenditures.[8] The results are shown in Figure 2.2 for the ten-year period from 1980 to 1989.

The data in the figure scatter considerably. Nevertheless, they have a correlation coefficient of 0.58 and the slope of the regression line indicates that about two dollars of sales were generated for each one dollar of R&D expenditure. And, since the square of the correlation coefficient is 0.34, about 1/3 of the sales increase can be associated with the R&D expenditures. It is

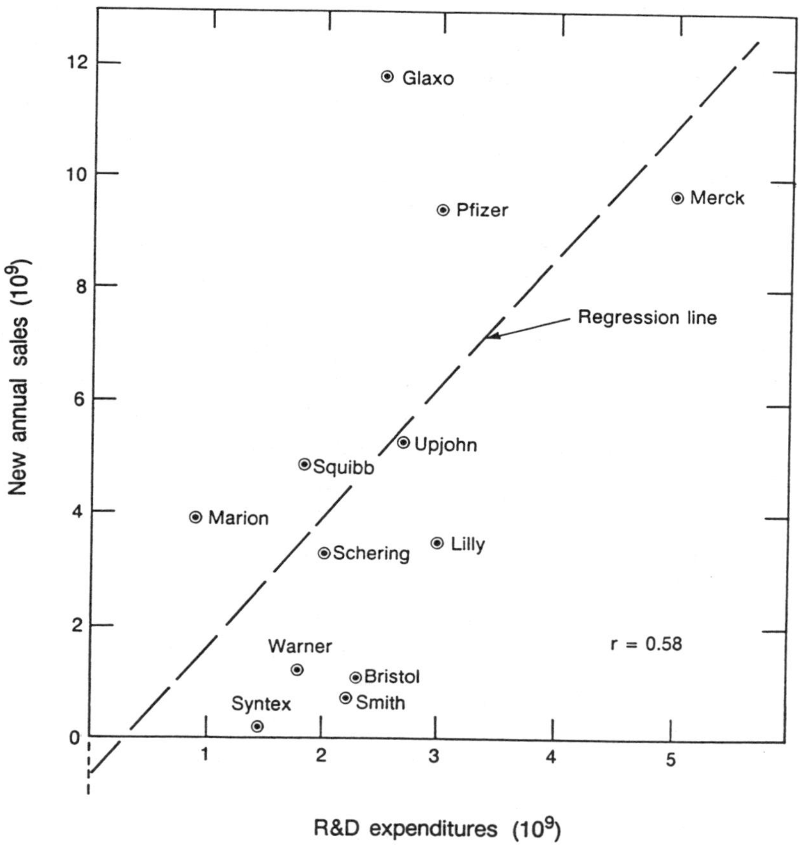

FIGURE 2.2. Sum of new annual sales versus sum of research expenditures for the major U.S. pharmaceutical companies. Period from 1980 to 1989. (Source: Teitelman and Baldo, "Grading R&D").

assumed here that R&D expenditure is the independent variable; the justification being common sense. It seems quite unlikely that increased sales today could cause R&D expenditures 10 to 15 years earlier.

EPRI Study

Another approach to the question of the value of research has been provided by a study that was done by the Electric Power Research Institute (EPRI).[9] This research institute is supported by a "tax" on 500-plus companies of the utility industry. This industry grew out of the Edison companies that were started more than a century ago. Needless to say, the Edison companies came from the research that was done at Edison's laboratory at Menlo Park, New Jersey. This was the first place in the world where industrial research was practiced on a large scale (more than 100 people) with the expressed interest of creating economically significant inventions.

One of the interesting features of the EPRI study is that it demonstrates that even a highly mature, highly regulated, commodity-based industry can enjoy a positive benefit/cost ratio from research activities.

In the study, 24 of EPRI's more active client companies were selected. The plan was to determine how their concrete benefits from research work done by EPRI compared with the costs to the clients (14 years of "taxes" paid to EPRI). The 24 companies were well distributed geographically. The self-interest of EPRI in the results was minimized by having the companies themselves responsible for executing the studies. Some 500 R&D "products" were identified: some used by as many as 17 of the participating companies; one used by only one. The benefit/cost ratios that were identified ranged from 1.2 to 7.6 among the 24 participants. Much of the variation is attributable to the diversity of methodology that arose within the individual companies; so the average ratio has no meaning, but it should be noted that the individual values were always greater than one.

The distribution of use of the "products" of the research projects are shown in Figure 2.3. The total number of use events was found to be 1,490 and they were distributed as indicated by the figure where the fraction of products used by the participating utilities decreases exponentially with the rank number of the company. Seventeen of the 24 companies used at least one product, 7 companies used none, while the most active company used 435 of the 500 products. The largest among the companies made the most use of the research results, probably because they could couple with the EPRI organization more effectively than the smaller ones. The sizes of the companies varied by a factor of 180 in terms of electric power output. It would be interesting to know whether the rank numbers in Figure 2.3 correlate with the sizes of the companies.

AN OPTIMAL SPENDING MODEL

Several years ago a more general approach to evaluating the benefit of research for industrial companies was proposed.[10] Companies seek to maximize their overall values. These are determined by the prices that are offered for them in the marketplace (stock market). One of the factors that determines the value of a company is the success, or lack thereof, of its

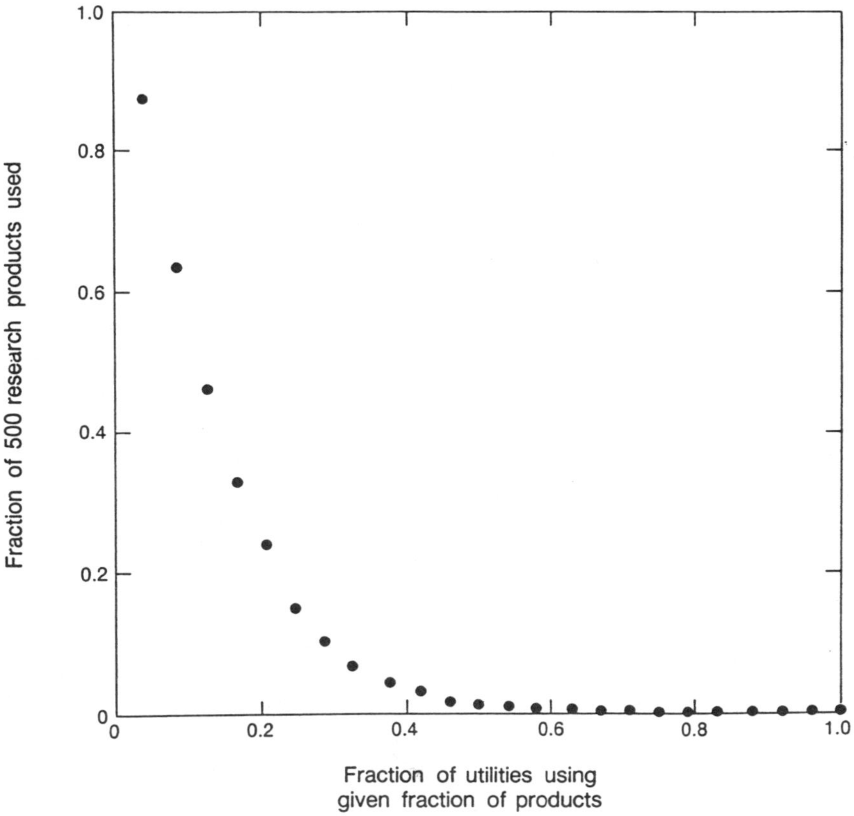

FIGURE 2.3. Distribution of 500 EPRI research products among 24 client utility companies. (Source: Whitaker, "Measuring the Value of R&D").

research program. Without some success in this area, it will have difficulty bringing new products to the marketplace. Since older products tend to become obsolete, this is an essential activity for many companies. For other companies, research may be needed to find ways to reduce manufacturing costs. For still others new marketing tools are needed. And so forth. If these needs are not met, the company may go out of business. The stock market expresses an opinion of which companies have better research programs than others through the price/earnings (P/E) ratio that is offered for their stock. Research success is not the only factor that determines the P/E ratio, but it is one factor.

The price/earnings ratio is an expression of the market's estimate of the trend of a company's earnings. If it is supposed that the trend is downward, the ratio that a buyer is willing to pay will be less than the market average for companies of a particular type. If the trend is upward, a buyer will be willing to pay a premium in the hope of benefiting from the future growth. Thus an estimate of future prospects is included in the many factors that determine P/E ratios. Since a strong research effort improves future prospects, it tends to increase the P/E ratio. The result is that the marketplace gives an informed and independent measure of research effectiveness; *vice versa*, research effort improves a company's market value.

Other financial parameters describe the past and present performances of companies, and proposals have been made to use them to estimate the effects of research spending. For example, return on sales, and sales growth rates have been extensively studied.[11] However, the one financial parameter that intrinsically includes a futures component is the P/E ratio. It helps, of course, if the market is "efficient." Short-term fads are aberrations; they exaggerate future expectations through inflated P/E ratios. An example was the biotechnology fad from about 1978 to about 1984. During this period P/E ratios were inefficiently (that is, irrationally) inflated to infinity for many small biotechnology companies even though they had no sales and hence no real earnings. Excluding such fads, the P/E ratios of companies that have strong R&D programs tend to be higher than those that do not.

A method for more directly assigning monetary value to research benefits would be desirable, but is not available. But most research benefits are not obtained until a considerable amount of time has passed after the corresponding expenditures have been made. Furthermore, the future benefits cannot be predicted effectively. Even if they could be, they might be aborted by future events. For example by competitive research results or by an ineffective commercialization program. Thus only an indirect method can provide partially objective information.

One of the better methods for making predictions is to use Delphic statistics. That is, to average the predictions of a large number of "oracles." If the oracles are informed, motivated, and independent, then significant, although approximate, results can be obtained.

Since the long-term future of an industrial company depends on its having an effective research program, and one of the functions of stock market analysts is to predict the future prospects of corporations, it seems reasonable to view them as oracles. That is, to use the collective opinions of the stock analysts as delphic statements although they may sometimes be flawed. Price/earnings ratios are a convenient means for doing this because one of the factors that determines them through the market analysts is future prospects.

An example of the correlation found between P/E ratios and research spending (S) is shown in Figure 2.4. The R&D expenditures are relative. That is, they are divided by the company's sales (S) for the same period in order to normalize them yielding the ratio, R/S. Although it might be argued that high P/E ratios lead to large R&D budgets by representing excess gross earnings that are available to be spent on research, this is certainly not true

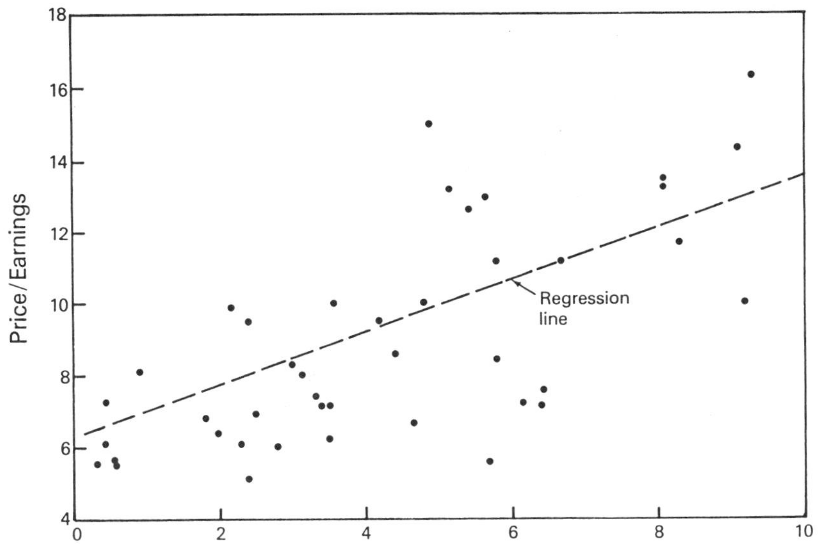

FIGURE 2.4. Price/earnings ratios versus relative R&D expenses for the 50 companies with the largest R&D expenditures (1981–82 averages). (Source: Wallin and Gilman, "Determining the Optimum Level for R&D Spending").

for small companies. Therefore, it seems far more sensible to believe that the R&D expenses are the independent variable, and the P/E ratios follow them.

In order to find out whether the P/E versus R/S (research spending) correlation is stable enough to provide a basis for a model, it was studied for a series of seven years (1976–82). It was found indeed to be reasonably stable with the average correlation coefficient being 0.59, and the range being from 0.48 to 0.66.

Another check of validity is to look at the companies in the top 10 percent of the distribution, and the bottom 10 percent. Companies with strong reputations for marketing new products based on research are in the top 10 percent. The converse is also true. The bottom 10 percent of the companies have poor reputations for marketing new products.

Given the correlation just described, a casual observer might think that the more a company spends an research, the better. This is not true, of course, because the research expenditures reduce the net earnings. Thus there must be an optimum amount to spend on research. It is determined by a balance between the benefits and the costs.

Figure 2.5 illustrates a simple model of this balance. The schematic diagram

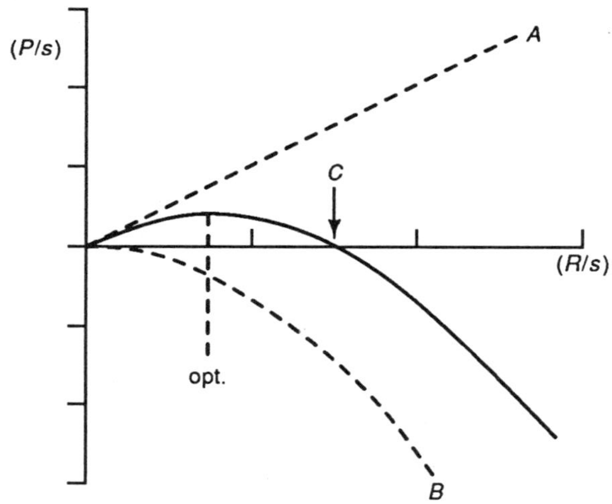

FIGURE 2.5. Schematic diagram illustrating model for determining optimum research spending. Stock price/sales (P/S) plotted versus research expenditures/sales (R/S). Line A is the statistical effect of research on stock price. Line B is the effect of research expenditures on net earnings. The optimum value of R/S is indicated. At point C further increases in R/S tend to decrease the value of the company. (Source: Wallin and Gilman, "Determining the Optimum Level for R&D Spending").

is a plot of the price of a company divided by its sales (price/sales ratio) against its research spending divided by its sales (relative research spending). The price/sales ratio is the P/E ratio times the E/S ratio where E is the after-tax earnings, and S is sales. If there are no earnings, the model breaks down. This case will be discussed separately later.

The dashed line marked *A* in the figure comes from the correlation line of Figure 2.4, whereas the dashed line marked *B* comes from the fact that research spending is an expense that is subtracted from gross earnings before taxes. Thus, while it benefits overall future value (that is, *P*), it also reduces immediate value because it reduces net earnings. Its net effect is indicated by the solid line in the figure. As research spending increases up to point *C*, the effect of the spending is positive, then it becomes negative. Also, the positive passes through a maximum indicating that there is an optimum amount of spending.

The mathematical derivation of this model is given in Appendix I. The feature that may be of most interest is the expression for the optimum research spending, R^*. It is as follows:

$$R^* = (1/2)\,[E_b - (S/K)(P/E)_o]$$

where E_b is the earnings before research spending and taxes; K is the slope of the regression line in Figure 2.4; *S* is sales; and $(P/E)_o$ is the intercept of the regression line in the same figure. It is the P/E ratio in the absence of any research spending.

It has been found[12] that the values of *K*, and therefore of R^*, vary with the industry group, and with the size of the company. Figure 2.6 illustrates this. It compares large and small chemical companies as well as large and small electronics–instrument companies. Large is defined here as over one billion dollars, and small is under 300 million dollars in annual sales. The figure indicates that for large chemical companies the optimum spending level is about 2 percent of sales, while it is about 5 percent for small chemical companies. There is an even larger difference for small versus large electronics–instrument companies.

The results given by this model are consistent with general behavior in industry. However, the quantitative aspects are not as precise as might be desired. To obtain better precision, a better means for correlating future values with present research spending would be needed, but it is not known what this might be.

The reader may have noticed that the taxation factor does not appear in the expression for the optimum research spending. This makes it doubtful that tax policy can have an effective influence on research spending levels.

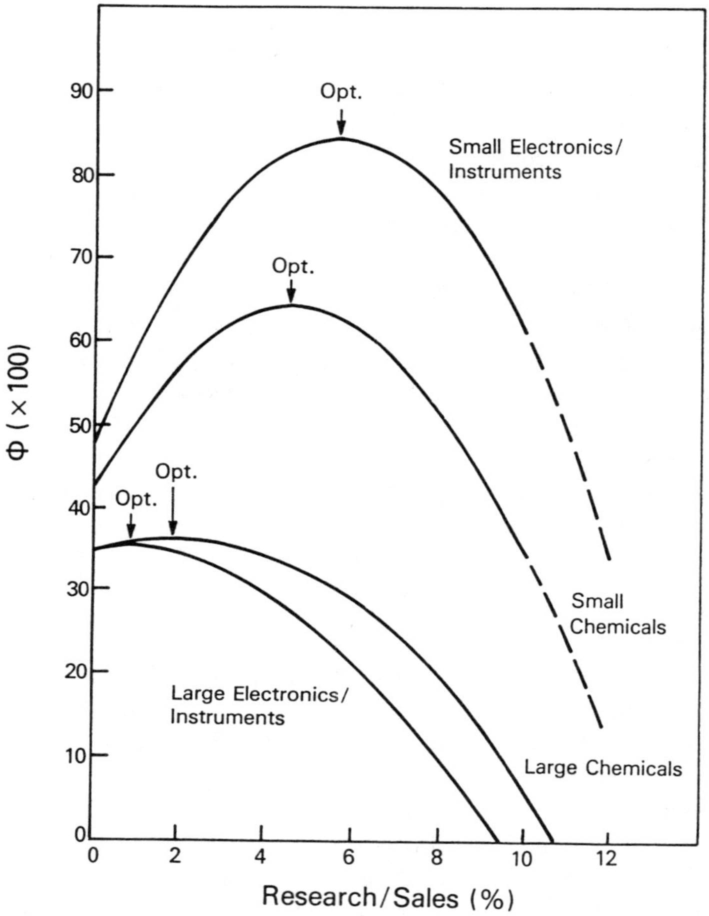

FIGURE 2.6. Effect of industry segment and company size on optimum research spending. (Source: Wallin and Gilman, "Determining the Optimum Level for R&D Spending").

As others have pointed out, tax breaks simply lead to relabeling expenditures in order to reduce the overall tax bills of companies.

PRICE/SALES RATIO CORRELATION

For established, stable companies P/E ratios are the best available indicators of future potential. But for new companies, and companies in a state of transition, the earnings may be small, zero, or negative, so the P/E ratio is meaningless. Although low earnings are a warning signal, companies that suffer from them may still be relatively rich in assets. For such companies, the ratio of the stock price to the annual sales per share is more meaningful. This has been discussed at length by Fisher.[12] Investors are willing to pay somewhat more for the current sales of companies that are expected to grow steadily in the future. The nominal norm for this ratio is unity, but it varies from as little as 0.1 to as much as 7.

Earnings are a more meaningful measure of effectiveness than sales for steady-state situations because earnings depend on how skilful a company is in selling its product while limiting costs. However, during transient situations such as startups, turnarounds, changes of direction, and the like, sales have meaning because they are a precondition to earnings.

Figure 2.7 shows P/S plotted against R/S where *P* is the total market price of the company, *S* is the annual sales, and *R* is the research expenditures. Data for 100 companies are shown.[13] The solid line through the data is the

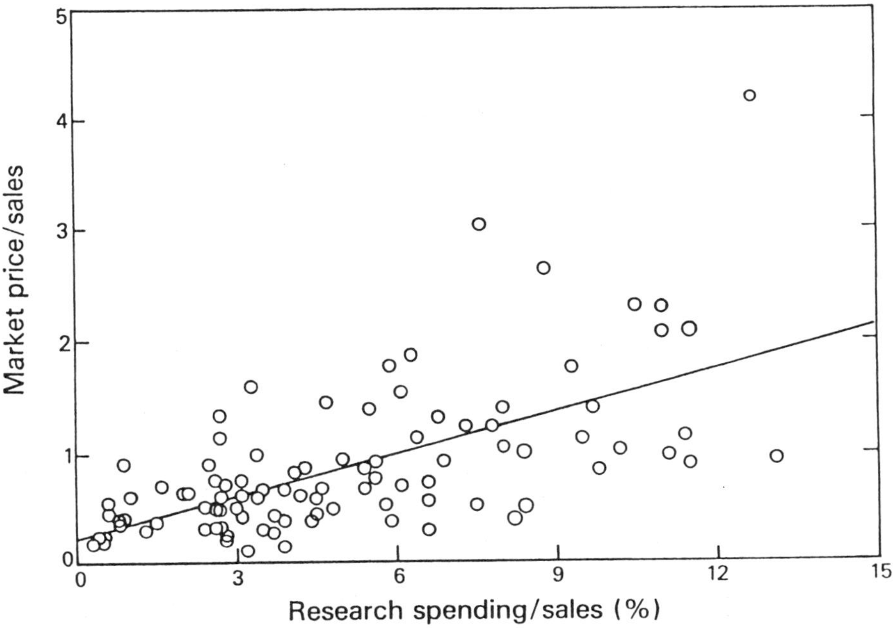

FIGURE 2.7. Correlation between the market price/sales ratios (P/S) of companies and the relative amounts they spent on research (R/S). The data are for the 100 companies that spent the most on R&D in 1983. (Source: Gilman, Inside R&D, vol. 13).

least squares regression line. Its slope is 12.5, and the correlation coefficient is 0.61.

A similar plot was made using the P/E ratios of the same 100 companies (with ten deletions of companies that had earnings deficits). For this plot the slope is 108, and the correlation coefficient is 0.54. The difference between this and 0.61 is probably not significant.

The slope coefficient (0.125) measures the statistical leverage that research spending exerts on a company's market value. Thus, an increase in R/S of unit amount might increase the P/S by 0.125. Hence, the statistical leverage is large. Nominally, doubling the relative research spending should (on average) increase the market value by 60 percent. But there is too much scatter for this to be realistic. A better interpretation is that if a given company were to increase its spending, and behave like those companies that spend twice as much on research, its market value could be expected to increase 60 percent.

In order to see whether there is qualitative consistency between the statistical pattern and the general reputations of companies, consider the short list of companies with the highest P/S ratios (greater than two). They are:

Hewlett-Packard	Wang Laboratories
Merck	Searle
Intel	Syntex
AMP	

Most of these are admired for their strong research programs.

On the other hand, consider the seven companies with the lowest P/S ratios (less than 0.25):

Ford	Gulf
U.S. Steel	International Harvester (Navistar)
Texaco	Mobil
Allied-Signal	

Few of these have notably strong research programs. Thus qualitative consistency is confirmed, although it is also clear from these lists that R/S is not the only factor that determines a company's P/S ratio. Other strong influences were acting on them in the 1983–84 period.

It should be emphasized that a plot like that of Figure 2.6 describes the behavior of a large ensemble of companies. It does not describe how individual companies will behave because too many factors are involved. Nevertheless, these data support the proposition that effective research is a lever of considerable power for companies to use in their quest for increased market value.

GENERAL POLICY

It is clear from the previous discussion that research activities can be justified within commercial firms in terms of economic factors that can be quantitatively defined. But what about nonprofit organizations. What about society as a whole? How much research should be done at the discretion of the researcher, and how much should be directed by higher authorities? If people and corporations spend some of their income directly on research, how much of their taxes should be used for the same purpose? None of these questions have pat answers. Only the ballot box with all its imperfections decides most of them at present.

Some perspective can be gained by considering research expenditures relative to the gross national product (GNP). In 1988 the GNP was about 5.4×10^{12} dollars. Since the bulk of the GNP consists of commodities, the optimal spending model suggests that the optimal fraction for R&D spending is about 2 percent or 110×10^9 dollars. Then, if the rule-of-tens is applied, about 11×10^9 dollars should be spent on research after 99×10^9 dollars have been spent on development. Thus the amount to be spent at the discretion of research workers themselves would be about 1.1×10^9 dollars. These numbers are nominally consistent with actual practice, except that "discretion" has a special meaning in these times of reviews piled upon reviews.

ASSET GENERATION

As stated previously, one of the better measures of the effectiveness of a business is its return on assets, or ROA. This is the surplus it generates from the assets it already possesses. It may be written:

$$dA/dt = rA$$

which states that the rate of increase of assets is proportional to the present assets with r being the proportionality coefficient. Note that r = ROA.

This expression is readily integrated to yield:

$$A(t) = A_o \exp(rt)$$

where A_o is the assets at the beginning of a period of time, $A(t)$ is the assets after time, t, has elapsed, and exp is the exponential function. This shows that $A(t)$ is very sensitive to the coefficient, r. For example, if r, the ROA, changes from 5%/yr to 15%/yr, then after ten years have passed, the assets will have increased roughly a factor of three more than they would have at the lower level of return.

New businesses created through research results, if successful, tend to have values of r in the 20 percent to 40 percent range, some even higher. Old businesses with products that have become commodities tend to have values in the 5 percent to 10 percent range. So it is easy to see what the motivation for doing industrial research is from the businessman's viewpoint.

The argument thus far has not taken the costs of getting research results into account. When this is done, the picture becomes much more complex because the return on investment must be considered. The cost of doing research (investment in research) and the resulting return (if any) must be compared with returns that can be made in other ways, such as investing in plant expansion, real estate, government bonds, and so on.

A priori comparative investment analysis is very likely to be misleading in the case of research because it almost invariably underestimates the returns of successful efforts, and overestimates those of unsuccessful ones—not because of faulty analysis, but because of faulty projections of the future. In fact, projections of the future are usually so poor, that analysis of research investments based on projections lead to the conclusion that they should not be made because there will be no return, or a smaller return than would be expected from a bank account. The reason for this is that the future is unknown. Even short-term linear projections are treacherous, as every stock trader knows.

The feature of the future that saves the day for research investments is its richness. If it were, in fact, simply a linear projection of the present, there would be little point in doing research because there would be no surprises. It is the synergisms, and the serendipities, and the accidental discoveries that provide the opportunities both for intellectual excitement and for new businesses. Nonlinearities are the order of the day in the future. Long live the kinks and cusps of the future!

It is the unpredictability in research work that provides its fascination. In this respect it is like sports. The outcome is not known until the end. "Its not over, until its over." This is why Barzun (1964) has called it the "glorious entertainment." Unlike sports, however, research opens up new opportunites by changing the rules. Branching goes on continually, which increases the scope exponentially. Someday we may come to the ends of the branches, but there is little evidence for this to date. If this happens, the expression for the rate of change of assets will become:

$$dA/dt = rA - sA^2$$

and, when A becomes equal to r/s, dA/dt becomes zero so A will no longer increase with increasing time. This can happen for a company, an industry, or a country. If $A(t)$ is known for span of time, and the situation to which it applies does not change, then a value for s can be derived. Once s has been determined, the time at which dA/dt becomes zero can be calculated. In other words, a prediction can be made long before A actually stops increasing.[14,15] This may allow adjustments to be made that will delay, or prevent, the stoppage of growth.

THE COMMERCIALIZATION CHAIN

The route to a new commercial product is like a chain, and therefore is vulnerable to weakness in one or more of the links. In other words, it consists of a series of sequential steps that are dependent on each other. Indeed the subsequent steps tend to need support from the preceding steps. This makes it exceedingly difficult to use case histories to establish cause-and-effect relationships between the research step and the point-of-sale step. For every case in which there is an apparent connection, another case can be cited in which the connection became blocked or broken. This is why "hindsight studies" have not led to self-consistent conclusions.

Inventions coming from research do not of themselves result in commercial products. They are necessary, but not sufficient. They are nuclei that may grow or may become stunted. If important, they will almost surely become transformed in some way before they become effective in a general marketplace. Thus there is no "germanium valley" because silicon transistors are far more effective than the original germanium inventions. Similarly, the first lasers used ruby rods for the active medium, but gallium arsenide has become much more important commercially.

The lack of close connection between invention and product has led to the misconception that serious striving in research may not be worthwhile because other companies may reap greater benefits than the original inventor.[16] This is sometimes true, of course, but not because the investment in research was bad. Almost always it is because there is weakness elsewhere in the chain. If the chain has good links, the inventing company becomes the principal producer and profits handsomely. In fact this is the best way for a company to obtain and hold the leading market share. The invention of nylon brought large benefits to the DuPont company, as did the invention of transistors to AT&T, the invention of instant photography to the Polaroid company, the invention of laser printers to the Canon company, and so on. A company like the Illinois Tool Works has enjoyed the benefits of its hundreds of inventions for many decades.

Just as research publications are distributed among authors according to Pareto's law (see Chapter 4), and patents are distributed among inventors according to the same law, so successful new products are distributed among companies. This strongly suggests that statistical factors (random pluses and minuses) determine the fates of inventions along their commercialization routes. If this is correct, the differences between any two specific cases will tend to be greater than their similarities. It follows that case history studies are unlikely to yield general principles.

NOTES

1. J. Barzun, *Science: The Glorious Entertainment* (New York: Harper & Row, 1964).
2. A. F. Burstall, *A History of Mechanical Engineering* (Cambridge, MA: The M.I.T. Press, 1965), p. 35.
3. W. Dunham, *Journey Through Genius*, (New York: J. Wiley & Sons, 1990).
4. S. Ramo, *The Management of Innovative Technological Companies* (New York: J. Wiley & Sons, 1980).
5. M. Josephson, *Edison—A Biography* (New York: McGraw-Hill Book Company, 1959).
6. L. A. Hawkins, *Adventure into the Unknown* (New York: William Morrow & Company, 1950).
7. D. A. Hounshell and J. K. Smith, *Science and Corporate Strategy—Du Pont R&D, 1902–1980* (New York: Cambridge University Press, 1988).
8. R. Teitelman and A. Baldo, "Grading R&D," *Financial World* Vol. 158 no. 2 (1989) p. 22.
9. R. Whitaker, "Measuring the Value of R&D," *EPRI Journal* Vol. 11 no. 6 (1986) p. 12.
10. C. C. Wallin and J. J. Gilman, "Determining the Optimum Level for R&D Spending," *Research Management* Vol. 29 (1986) p.19; see also, *ibid.*, 21, (1978) p. 34.
11. L. W. Ellis, *The Financial Side of Industrial Research Management* (New York: J. Wiley & Sons, 1984) Chapter 8.

12. K. L. Fisher, *Superstocks* (Homewood, IL: Dow Jones-Irwin), 1984.
13. J. J. Gilman, the sample is the list of 100 U.S. companies that spent the most on R&D in 1983 (*Inside R&D*, Vol. 13 no. 23, 1984). The prices per share came from the January 4, 1984, issue of the *Wall Street Journal.* The shares per company came from the 1984 edition of *Standard and Poor's Stock Guide.* Comparative P/E ratios were also for January 4, 1984.
14. C. Marchetti, "10^{12}: A Check on the Earth-Carrying Capacity for Man," *Energy*, Vol. 4 (1979) p. 1107.
15. T. H. Lee and N. Nakicenovic, "Technology Life-Cycles and Business Decisions," *Int. J. Technology Management* Vol. 3 (1988) p. 411.
16. L. R. Cohen and R. G. Noll, *The Technology Pork Barrel* (Washington, D. C.: The Brookings Institution, 1991) Chapter 2.

3

WHAT IS WORTH WORKING ON?

It is much more exciting not to catch a big fish than not to catch a little fish.

—Albert Szent-Gyorgyi,
Nobel Laureate in biochemistry

Opportunity is difficult to generate on demand. There is much to be said for waiting for it to knock. However, activity is expected by the sponsors of a research group even when the group has nothing that is really worthwhile to work on. Anxiety ensues. It is often alleviated by making rash promises or by retreating to activities designed to cut costs. It would be better to recognize that "opportunity is where you find it," not where one might wish it to be, and wait patiently while strengthening the organization in anticipation. This can be done with confidence that an opportunity will soon arrive. For diligent and reasonably clever research people it virtually always has.

What constitutes an opportunity? This question is unanswerable. The best that one can do is state some principles, or guidelines, that attempt to define what is not, or is unlikely to be, an opportunity. In retrospect, clear images of opportunity are easily generated, but not in prospect. The opportunities created by significant inventions are so much greater than those given by improvement that the discussion here will center almost entirely on inventions. This is in contrast to the majority of the research management literature that focuses on improvements to existing products and processes.

Exploration is the most likely source of major opportunities. Least likely is to follow some other explorer. In a nearly full parking lot, this latter may be like following another car when looking for a parking slot! On the other hand, if the parking lot is nearly empty, this might be a good tactic; especially if the lot is large. And, it is true that inventions often cluster around special situations.

Montroll and Shuler[1] have argued quite persuasively that dimensionless numbers made up of combinations of various parameters are often associated with major inventions. More inventions follow the primary one as the component parameters of the appropriate dimensionless number are varied in ways that are consistent with the dimensionless combination.

As an example they discuss the history of hydrodynamic devices. The behavior of these (ships, airplanes, windmills, harbors, etcetera.) is described by a differential equation that is not linear. Therefore, the behavior of a small ship is not simply proportional to that of a large ship and *vice versa*. For a given speed, long ships require less propulsive power than short ones. The difference can be calculated with the aid of a dimensionless number known as the Froude number (a combination of velocity, length, force, and density). By holding this number constant, tests of models allow for predictions of the behavior of full-scale ships. Thus the invention of new designs is greatly facilitated. Other dimensionless numbers can be used to aid the design of the sails, the propellers, the blades of the steam turbine, the ship's structure, and so on.

Montroll and Shuler point out that if the number of dimensionless quantities that is needed to describe a particular device becomes too large, designing workable embodiments of it may become nearly impossible. They suggest, for example, that this may be the case with controlled fusion reactors.

Stock picking has many corollaries with picking research topics. Books on the factors that lead to successful investing, such as that of Lynch,[2] confirm this. Both involve forecasting the future, selecting the right people, estimating future demand for a product, analyzing the economic environment, and other parallels.

Although there is no substitute for individual inventors with imagination, skill, and "lucky hands;" they are limited by a variety of constraints. Prudence requires an awareness of these constraints. For example, man-powered flight was not possible until materials with adequate specific strength became available. The incandescent light bulb could not have preceded the development of good vacumn pumps. Purification chemistry applied to semiconductors necessarily preceded transistors. And so on. In addition to these technical constraints there are others related to the current needs of a society. That is, to economic factors.

Another way of expressing the content of the previous paragraph is to say that inventors who work in harmony with their habitats are more likely to be successful than those who work otherwise. If the constraints are minimized, opportunities become maximized, and *vice versa*.

There are various ways of maximizing opportunities systematically, and other ways that require the mind of genius. We shall discuss some of the former. These include: demand charts, congruence, price trends, the market-share law, market-diffusion theory, and perhaps others that have not been recognized.

DEMAND CHARTS

All merchants understand that price and sales volume are inversely related. Most also understand intuitively, if not explicitly, that the relationship is nonlinear so revenues (price x volume) also depend on price.

At the height of the biotechnology craze (circa 1980), I inherited the responsibility for a substantial research program in this area. The program included the following topics: the extraction of specific minerals from complex ores using microorganisms; the production of oils by means of genetically engineered algae; the large-scale oxidation of an industrial organic chemical; photosynthesis studies; and DNA sequence detection through a hybridization analysis technique. An immediate question was: which of these should I continue to support, and which not? At the time, the government and financial communities were supporting all such topics almost indiscriminately.

In some cases comparative cost analysis could be applied. For example, consider the oxidation process. It illustrates the importance of examining a process in an overall way, not just part of it. In order to justify their proposition that biotechnology could compete with the conventional oxidation process, the biotechnologists used substrate–enzyme reaction rates as their process rates. Under good conditions these rates can be very high; certainly as high as many rates determined by inorganic catalysts. However, they ignored the need to transport the substrate(s) to the enzyme and the product(s) away. Even in a cell-free system, these diffusional processes limit the overall process rate. And in processes that use cells, they cause the overall rates to be two to three orders of magnitude lower than typical rates for commercial industrial–organic processes.

Putting aside specific problems of this kind, assertions were being made that virtually all chemical processing could be done more effectively using biological organisms than by conventional methods. Soon to be revolutionized were agriculture, mining, energy production, food processing, and medical practice. This caused a stampede of unwary investors similar to the gold rushes of former times. But it made contrarians nervous, including me.

Conveniently enough, the enthusiasts prevailed upon the Congress of the United States through its Office of Technology Assessment to sponsor a study.[3] The report of this study lists estimates of both the unit prices and the consumptions of a large number of existing and proposed products of biotechnology. Since the estimates were collected by enthusiasts, one could have some confidence that they were not underestimates. It was a simple job to plot the data in the form of a demand chart; namely, the logarithm of economic utility (price) versus the logarithm of consumption.

Figure 3.1 shows prices ($/#) plotted against consumptions (#/year) for a wide variety of organic chemicals with prices varying from about 0.015 to 5.0 $/#, and consumptions ranging over about 6.5 orders of magnitude. The line drawn through the data points is the least squares regression line. Although some of the data points lie substantially off the regression line, the overall correlation of the data is high with a correlation coefficient of −0.90.

How many of these chemicals might be produced economically through bioprocessing? Although there are some in vitro bioprocessing methods, fermentation is the most common technique. The retail cost of fermentation in the units of the figure was about $1/# in 1982. A horizontal line at the zero level of the figure can represent this since the plotted prices are retail. Most of the plotted points lie below this line. Thus most of the chemicals on the chart cannot be produced at competitive prices by fermentation.

It might be argued that the cost of fermentation could be substantially reduced through "scale-up." However, this is unlikely for at least one technical reason and for at least one fundamental reason. The cost of fermentation is relatively high because the process throughput is relatively low (by two to three orders of magnitude compared with conventional industrial–organic processes). By concentrating the broth, the throughput can be improved, but the viscosity of the broth increases nonlinearly with concentration so this approach is quite limited because the broth soon becomes so viscous that it cannot be stirred effectively. The fundamental limitation in fermentation, as shown by measurements of what is known as the Thiele modulus, is the rate of diffusion into and out of cells. The nutrients must diffuse in, and in most cases the products must diffuse out because destruction of the cells is too expensive. This contrasts with most industrial–chemical processes in which convection limits the rate. Convection is much faster than diffusion.

For simple reactions, some of the difficulties of fermentation can be avoided by extracting enzymes and using them as process catalysts. But these

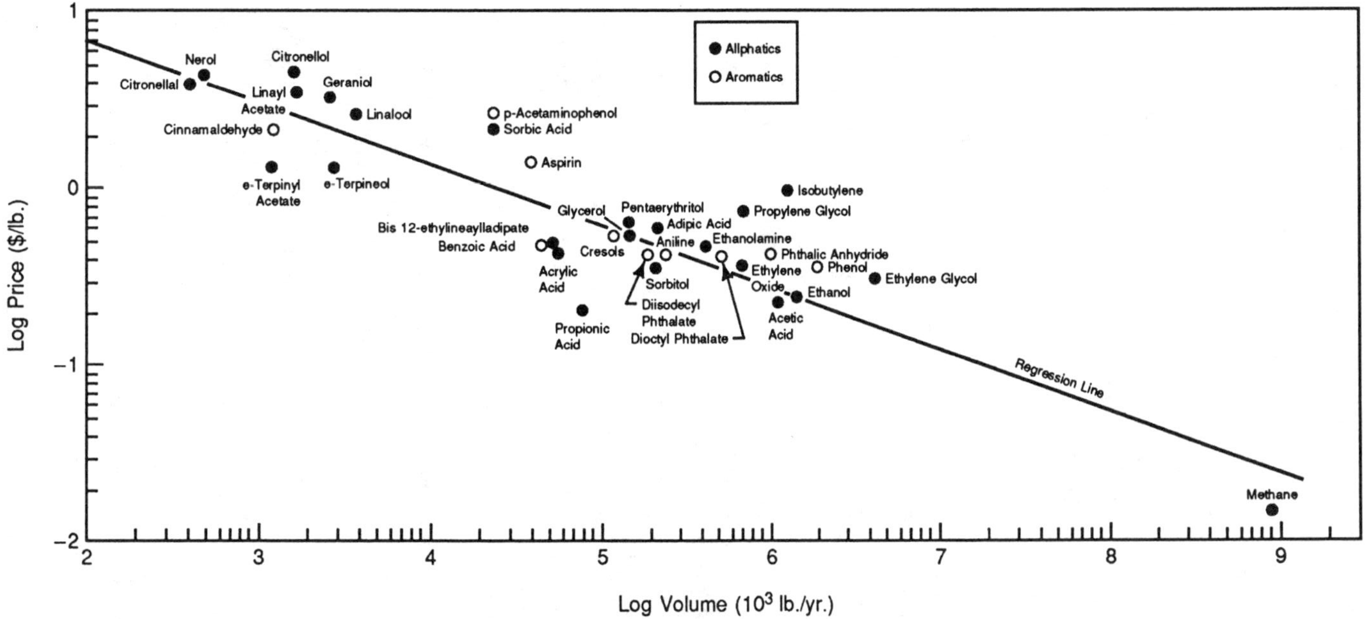

FIGURE 3.1. Price/volume data for selected organic chemicals.

are the exceptions in biotechnical production processes. Often several enzymes are needed before the end product is reached. They must operate in series so there is little or no advantage in using them in vitro instead of in the highly organized environment inside a cell.

Since $/# times #/yr. yields $/yr., Figure 3.1 is also a revenue plot as was indicated previously. The intersection of the fermentation cost line with the regression line occurs at the 20×10^6 $/yr. revenue point. Thus no single chemical bioproduct could be expected to be the basis of a large business, usually defined as over 100×10^6 $/yr.

Consideration of these relatively straightforward points led me to the conclusion that a realistic research strategy in the biotechnology area had to be considerably more sophisticated than most people were contemplating at the time. This has been confirmed by subsequent events. It also led to the realization that demand charts are not only useful for commercial planning, but also for research planning.[4] Not just for deciding yea or nay regarding proposed projects, but also for choosing between various potential tactics. In addition, there appears to be a place for demand charts when considering national technology policies.

An example of how attention paid to demand charts might improve policy selections is the case of titanium metal. In the decade centered on 1950, it was repeatedly announced that titanium was the new wonder metal that would replace much of the consumption of aluminum and stainless steel in both the civilian and military sectors. Hundreds of millions of dollars were poured into R&D—mostly development. Advances were certainly made, but not commensurate with the expenditures. This can be seen by examining of Figure 3.2, a demand chart for structural metals.

Titanium ore is relatively cheap, but extracting and fabricating the metal is expensive. At its relatively high price, the point on the chart corresponding to titanium is at a revenue level of about 800 million dollars per year, and most of its consumption was for military applications. This may be compared with 60 billion dollars per year for aluminum. Given the close correlation of the various metals with the regression line, it should have been clear to planners that titanium would always be of relatively minor commercial importance because of its price relative to other metals.

A more recent example is that of oxide superconductors. Estimates of their manufacturing costs indicate relatively high prices.[5] Therefore, they cannot be expected to be used on a massive scale, indicating that prudence should be used in expending research resources on them. The actual spending spree that began in 1986 is likely to end in a painful hangover.

Demand charts must have suitable coordinates if they are to be useful. The

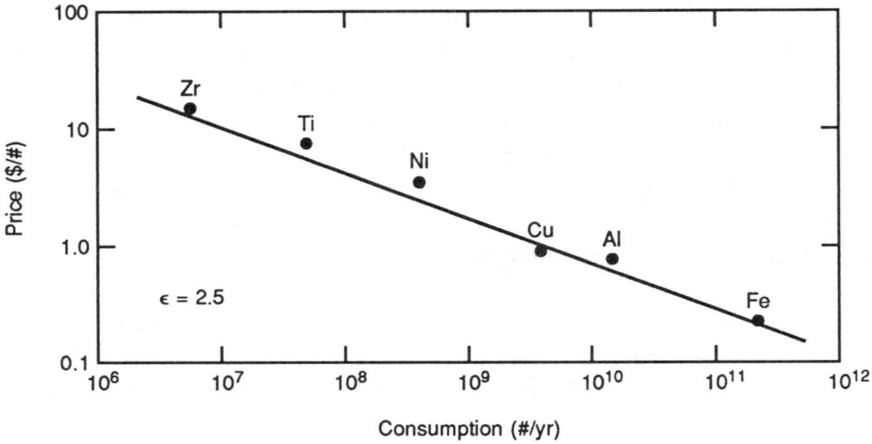

FIGURE 3.2. Price and consumption for the principal structural metals in the United States in 1980. (Source Engineering and Mining Journal, 1981).

most common measure of the product that is used is its weight. Its volume is also often used. These measures are fine for commodities where the utility of the product can be closely equated with its quantity, but when the utility has a more subtle connection to the product, other measures are likely to be more appropriate. For example, for electric batteries the price per unit and the number of units consumed per year do not correlate very well. But the correlation is excellent between the price per unit (KW-hr.) of electricity stored and the amount of electricity consumed per year for a wide variety of battery types.

It is also important to compare similar forms. Comparing sand with silica laboratory ware, carbon electrodes with diamond, or canned asparagus with that served in a 3-star restaurant will not lead to sensible correlations. Since increases in utility determine the commercial values of inventions, estimating this parameter is of utmost importance, even though it is often an elusive parameter to estimate.

On Consensus Versus Contrarianism

Anyone familiar with the idiosyncracies of stock markets knows that a consensus opinion about what a market will do next is a poor predictor. In fact the anticorrelation between the consensus opinion and the actual behavior is significantly better than the correlation. When a majority of the stock analysts predicts upward price movement, the probability that downward movement will actually occur is greater than one-half. And *vice-versa*. So much for consensus as a predictor. The opinions of experienced, educated, individuals are much better, albeit not perfect. The use of "experienced" in this context means someone who has a track record that can be compared with actuality.

In development work where the goals are well-defined, and in judging a completed piece of research or development work, a consensus of peers is quite likely to be accurate. Thus a consensus is good in providing criticism but not in providing a creative judgment. This is the problem in fact. Since a consensus does a good critical job, it is inappropriate for judging the unknown. Such judgments are best made by suspending criticism as much as possible and emphasizing optimism.

Many, if not all, major inventive advances have been discouraged by the consensus of opinion just prior to their time and often for a period of time thereafter. This has been true of railways, incandescent lamps, automobiles, airplanes, rockets, telephones, computers, transistors, lasers, and many more. A few enthusiasts drove these things into existence. Just as in the case of the stock markets previously cited, the antimorph also seems to operate at least part of the time. Consider, for example, the several energy technologies that the consensus approved of in recent years: solar photovoltaics, windmills, coal liquefaction, geothermal, magneto–hydrodynamics, inertially confined fusion, and so on. None of these has had a significant impact on the production of energy. So it does not appear to be the case that judgment by consensus is an approach with more than a modicum of reliability. The reason why it is unreliable appears to be that politics prevails over insight in reaching a consensus about the future.

THE MARKET SHARE RULE

With the well-worn image in mind that "the pioneer is the one with the arrows in his back," many research planners suggest that it is better to be second in a marketplace than first. There are examples in which this appears to be true. However, this idea should be viewed with caution in the light of the market-share rule. This relates market shares to the numbers of companies participating in a given market, and to the rank of a given company in the same market (that is, whether it is first, second, third, and so on). It is consistent with observed behavior for a variety of industries and products. Among these are: automobiles, steel, oil, breakfast cereals, specialty chemicals, Portland cement, New York newspapers, oil refining, aluminum, synthetic rubber, electric batteries, and so on.[6]

The distribution function for market shares is known in probability theory as the "partition of a magnitude";[7] or more descriptively as the "broken stick distribution."[8]

Suppose you sit down with a bucket of sticks having equal lengths and proceed to break each of them into two pieces at random places along their lengths. After you have broken a large number of them, what will be the average lengths of the individual pieces? Off hand, you might say one-half of the original length, but this would be wrong. The same thing is true of markets. If a market is broken into two parts, the participants do not share it equally. One participant gets 75 percent, and the other 25 percent, on average.

If a stick is broken at some random place, there is only one way to break it into two equal pieces, but there are two ways to break it into other proportions (say a 3/8th and a 5/8th piece). If the break is in the left half, the average place for it will be at half of that half; the same being true for the right half. Thus the average for the whole length will be 1/4 and 3/4 pieces. This assumes that special constraints such as patent agreements, or tariffs, are not present.

The analytic description of the rule for any number of participants is presented in Appendix II. The results of it for sets of up to ten companies is shown in Figure 3.3. For variously sized sets of companies, this figure shows how their market shares depend on their ranks. For example, for a set of three companies, the leading one is expected to have a 61 percent share, number two a 28 percent share, and the third an 11 percent share. See Appendix II for a graphic representation of this.

The justification for the numbers of the figure is simply that they are empirically observed for a variety of markets. The market must be reasonably homogeneous and stable, of course. It cannot be a very new one, or a high-fashion one that is highly volatile. An example of the operation of the rule is shown in Figure 3.4, which presents recent data for the large household appliance market (refrigerators, stoves, air conditioners etcetera).[9]

Another example is the battery industry. This includes alkalines, nickel–cadmiums, carbon–zincs, and small lead–acids. For the four largest participants in each of the American, European, and Japanese markets, the observed shares in 1984 were 52, 26, 12, and 8 percent, while the model rule gives 52, 27, 15, and 6 percent. Thus there is excellent agreement.

The pattern of market shares that the chart in Figure 3.3 displays suggests various strategic opportunities and traps. The first thing that can be seen is

NUMBER OF COMPANIES	RANK OF COMPANY									
	1	2	3	4	5	6	7	8	9	10
1	100									
2	75	25								
3	61	28	11							
4	52	27	15	6						
5	46	26	15	9	4					
6	41	24	16	10	6	3				
7	37	23	16	11	7	4	2			
8	34	22	15	11	8	5	3	2		
9	31	20	15	11	9	6	4	3	1	
10	29	19	14	11	9	7	5	3	2	1

FIGURE 3.3. Market Shares (%) for sets of up to ten companies ranked by share size in a particular market. These are the shares expected for a steady marketplace without significant transients. See Appendix II for more details.

that the companies in the first rank can increase their shares substantially by reducing the number of companies in the set. For example, by reducing the set to three, the leader in a set of four can increase its share 17 percent after the marketplace readjusts. And it only needs to buy up, or otherwise eliminate, 6 percent of its market.

For the companies of the middle ranks, however, the effect of reducing the set size is just the opposite. Their shares decrease or remain constant if the set size decreases. Thus there is a large incentive for market leaders to consolidate a marketplace by eliminating the companies with the smallest market shares. The corollary of this is that there is also a large incentive for the market leader to prevent an increase in the size of the set of participants. The leader loses 25 percent if the set size increases from one to two; if the set size increases from nine to ten, the leader loses 6 percent. There are corresponding large losses for the cases in between these two extremes.

By contrast, the lowly ranked companies gain if the number of participants increases providing they maintain their rank. If the set size decreases, however, they lose. Suppose, for example, that the set consists of eight companies, and those ranked seven and eight combine. They start with a total share of 5 percent, and finish with only 4 percent. Not much incentive there!

For most of the table mergers between horizontal nearest neighbors is not advantageous except for the first two ranks. This is one reason why real

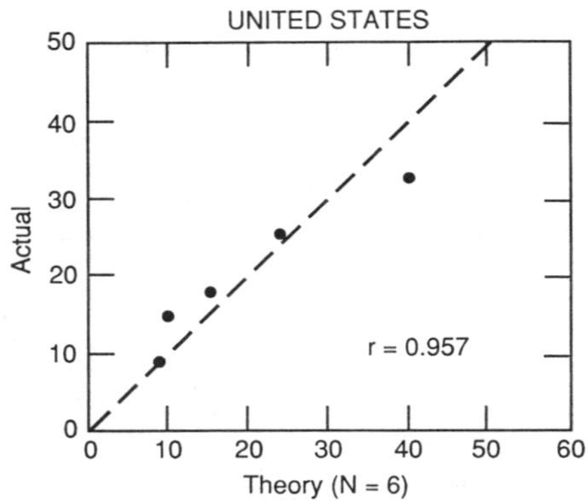

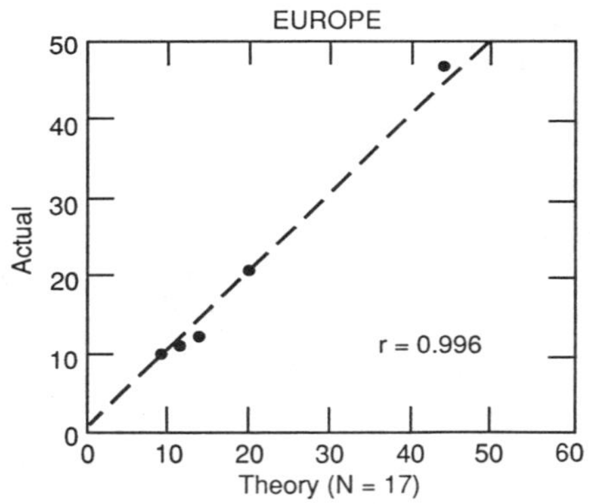

FIGURE 3.4. Comparison of actual (observed) 1989 market shares for large household appliances with those given by the market share rule. Data from Fortune, 121, 110 (1990). In each case N is the number of participants that gives the best agreement between observed data and the statistical rule. The diagonal lines in the figure indicate perfect agreement. The r values are correlation coefficients with r = 1 indicating perfect correlation.

mergers of this type are often not successful. Postmerger they immediately tend to lose market share to their more highly ranked competitors.

It is interesting that the companies in the second and third ranks of the table tend to be quite stable against changes in the number of participants in their marketplaces. This may be the basis of the idea that it is better to be second than first.

From the R&D planning viewpoint, trying to enter a marketplace from the bottom ranks, followed by working up through the ranks is folly. The only approach that makes sense is to aim for the top.

From a corporate strategy viewpoint, a distinct advantage of growing through R&D is that, when successful, it allows a company to start in rank one where there are powerful advantages even if several companies are participating in the same market. The best of all worlds remains that of being a monopoly (one company—100 percent share). Since research enables a company to introduce new or improved products to the marketplace, it can substantially help in preventing the number of competitors from growing.

The table in Figure 3.3 represents a statistical distribution that results from a balance between randomizing factors and organizing factors. This distribution is observed empirically, and the broken stick analogy indicates that it is consistent with common sense.

It may be noted that along the diagonal of the figure the market shares follow a Pareto distribution with a coefficient of two. That is, if the rank index is j, and the number of companies participating is j, then the market share is simply $1/j^2$. It is independent of the number of companies in the set. Thus there is a similarity between the distribution of market shares and other distributions such as the distribution of incomes or city sizes.

The market-share rule is very useful for testing hypothetical business plans. It allows a quantitative plan to be examined for sensitivity to changes in the number of companies that are participating in a market. This can be done with improved realism because the market-share rule limits the range of the assumptions that can be made.

RELATIVE PRICE TRENDS

Long before it was a generally recognized trend, the shift of the U.S. economy from manufacturing to service was reflected in relative price changes. A study of these changes by W. H. Fisher[10] indicated that the trend existed at least as far back as 1953. That is, about 25 years before it was discussed in the general press. Thus some guidance about the future is contained in relative price data. These trends alone cannot identify promising areas for research, but they can help, and they are objective.

Relative prices trends are those that exist in various sectors of the economy after correction for general inflation. They are the changes that occur in one industrial sector as compared with another. For convenience, Fisher adjusted the trends in the individual sectors in order to make their sum equal zero (in dollars corrected for inflation to a base year—1960). Figure 3.5 illustrates trends for a few sectors.

Price trends are an indication of the relative values that a society places on products. Positive trends indicate increasing demand (relative to other items), and *vice versa*. Some care must be used in interpretating these trends, of course, because other factors may cause relative trends. For example, terminal exhaustion of a resource will create a relative increase in its price. In the 1970s, the sudden change in the relative price of crude oil caused many instabilities in the relationships between prices.

Changes in relative prices serve as indicators of changes in a society's needs because the anticipation of such changes influence business success to an important extent. Success correlates well with filling the needs. Price trends that lead to cheaper raw materials (or components) in the future, and indicate an increasing demand (higher prices) for products made from the cheaper raw materials will enhance the climate in which a given business operates. This is illustrated in Figure 3.6, which is only meant to be schematic and therefore is extreme. More realistically, both raw materials and products would follow similar trends, but would converge or diverge.

Success is not ensured by operating in an area that has favorable price

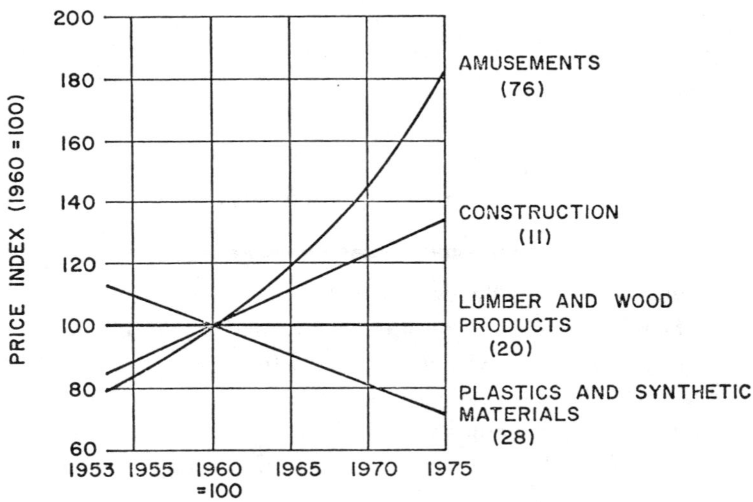

FIGURE 3.5. Sample relative price trends for four sectors of the U.S. economy for the period 1953–75. (Source: Fisher, "The Anatomy of Inflamon").

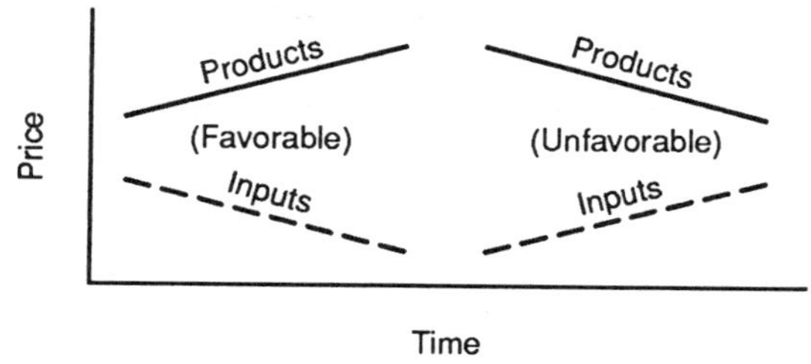

FIGURE 3.6. Schematic price trends illustrating favorable and unfavorable cases.

trends, but it does improve the probability. It increases the spread between input and output prices, thereby making larger margins possible. The efficiency with which the input materials are converted into saleable products will determine how much of the value added is retained by the company. For example, Figure 3.5 indicates that the input–output trend is favorable for converting plastics into objects used for amusements. On the other hand, the trend for converting lumber and wood products into plastics is unfavorable.

In addition to price trends, other factors are important to business opportunities. One of these is the rate of change of volume (growth rate). Others are factors associated with competition, the cost of money, and production costs. Nevertheless, some amount of objective guidance can be obtained by studying price trends and interpreting their meanings. The prediction of poor prospects from unfavorable price trends is rather unambiguous. When the input–output trends converge to nought, it becomes impossible to continue a business. The converse is that favorable price trends are a necessary, but not sufficient, condition for a high level of business success.

At the time of Fisher's first study (1971) the five industrial sectors with the fastest rising relative prices, and the five with the fastest falling ones, were the following:

RISING	*FALLING*
Forestry and fishery products	Textiles and floor coverings
Business services	Service industry machinery
Research and development	Household appliances
Amusements	Coal mining
Medical and educational organizations	Plastics and synthetic materials

More than a dozen years later, I commisioned him to do another study, which was completed in 1984. The discontinuity in oil prices caused by the 1972 OPEC event caused either small or large discontinuities in the relative prices of a majority of the other industrial sectors. These discontinuitites dramatically illustrate how severely the world economy was disrupted by the sudden change in oil prices. However, trends before and after the discontinuities tended to have the same slopes. The five fastest rising and falling sectors were now:

RISING	*FALLING*
Stone and clay products	Food products
Forestry and fishery products	Food and feed grains
Hotels and lodging places	Textile mill products
Construction, mining & oilfield machinery	Radio and TV receiving sets
Aircraft	Sawmills and planing mills

The rising set appears to reflect the influence of rising labor costs together with increasing relative demand, while the falling set appears to reflect relative slackening of demand for these products.

OPPORTUNISM

Price considerations and other forms of financial planning only provide guidelines for deciding what to work on, or equally important, what to stop working on. As often as not, invention comes through unplanned discovery while the researcher is studying something on a parallel but different track. Some examples are: the phonograph by Edison; penicillin by Fleming; the point-contact transistor by Bardeen, Brattain, and Shockley; radioactivity by Becquerel; aspirin by Hoffmann; polycarbonates by Fox; and many more.[11]

In all of the examples cited, as well as the sixty or more in the book just cited, *Serendipity* the "accident" that initiated the discovery did not occur at random, but to someone skilled in an appropriate art, who was an unprejudiced observer, and who was not under great pressure to complete a specific task. In other words the preprogramming of the project was minimal. But—the preprogramming of the motivation was maximal.

In every case the inventor was highly motivated to discover either something of scientific interest, or to find the basis of a practical invention. Herein lies the most important principle of all for finding worthwhile areas to work in. Depend on self-motivation! Find likely people to use the ideas of Chapter 4, give them psychological warmth and good tools, and let them ferment!

NOTES

1. E. W. Montroll and K. E. Shuler, "Dynamics of Technological Evolution—Random-Walk Model for the Research Enterprise," *Proc. Nat. Acad. Sci.*, Vol. 76 (1979) p. 6030.
2. P. Lynch, *One Up on Wall Street* (New York: Simon & Schuster, 1989).
3. J. C. Lashof, G. Kolsrud and Z. Harsanyi, "Impacts of Applied Genetics," Office of Technology Assessment, Congress of the United States, Library of Congress Cat. Card #81-600046 (1981).
4. J. J. Gilman, "Planning with Demand Charts," *Research Management*, Vol. 30 (1987) p. 32.
5. J. J. Gilman, "The Commercial Potential of High-Tc Superconductors," *Physics Today*, Vol. 41, (1988) p. 85.
6. J. E. Cohen, *A Model of Simple Competition* (Cambridge, MA: Harvard University Press, 1966).
7. W. A. Whitworth, *Choice and Chance*, 5th ed. (New York: Hafner Publishing Co., 1901; reprinted, 1951) Chapter 10, p. 205-8.
8. F. Mosteller, *Fifty Challenging Problems in Probability with Solutions* p. 63, (Reading, MA: Addison-Wesley Publishing Company, 1965) p. 63; Dover Edition (New York: Dover Publications, 1987).
9. Fortune, Vol. 121 (1990) p. 110.
10. W. H. Fisher, "The Anatomy of Inflation: 1953–1975," *Scientific American* Vol. 225 (1971) p. 15.
11. R. M. Roberts, *Serendipity—Accidental Discoveries in-Science* (New York: J. Wiley & Sons, 1989).

4

WHO CAN DO IT SUCCESSFULLY?

The reasonable person adapts himself to the world; the unreasonable one persists in trying to adapt the world to himself. Therefore, all progress depends upon the unreasonable man.

—George Bernard Shaw

All organizations that aspire to the production of creative work need to employ people with special qualities. If these qualities were measurable the task of assembling a research organization would be greatly simplified. But they are not. And they include enormous scope in terms of skills and interests and values. In addition, the profiles of characteristics of these people tend to be quite different from those of people in the general population.

Proof of the assertions in the previous paragraph is easily had. Here is the simple prescription. Sit in the lobby of a convention hotel during various conventions. Choose a few differing social groups first. For example, begin with a convention of financial executives. Then, one of industrial plant managers, and later one of advertising designers. The typical characteristics will be distinctly different for each group of people; including their clothes, hair styles, gestures, and speech patterns. And, if their external appearances and mannerisms differ, imagine how their thought profiles must vary from one to another!

Now sit in the same lobby while various conventions of technical people take place. Let them be conventions of physicists, civil engineers, and surgeons. Each group will be readily distinguished.

Next, observe people within research organizations who perform different functions. Consider chemical companies. Compare the hotel lobby when the practicing chemists are convened with when their research vice-presidents are meeting. There will be little resemblance.

Thus, people who practice various professions have personas associated with themselves. These are not rigorously defined, but they are easily differentiated. Intimately associated with these personas are skill profiles; that is, their talents.

There are two aspects to the selection of talented people that are most important. One is the nature of the talent, and the other is its intensity. In order for a research organization to be competitive it must be able to select people with particular talents at the highest possible intensity levels. Evidence for this will be given shortly.

The intensity with which a talent is used depends, of course, on the motivations of the talented person.

The principal purpose of doing industrial research is to effect synthesis. That is to create something new by bringing together a variety of factors in some new configuration. This may result in a new philosophy, a new device, a new medicine, a new machine, a new measurement, a new engineering system, a new social system, and so on. To do this requires that a person carry out a sequence of operations, including preparing or accumulating information in a specific topical area, selecting the most pertinent factors, executing projects to develop these factors, discovering a new combination, confirming the discovery, and communicating the new result. Doing just one of these operations is not enough; nor is it enough to do most of the sequence with one, or more, steps omitted. It might be thought that if a given person does a few of the first steps someone else would do the rest. But this is almost never the case for various psychological and social reasons.

People who are capable and motivated to execute a complete sequence of demanding tasks are very rare, and they become increasingly rare with the length of the sequence. Suppose there are six operations in a sequence that must be completed successfully in order to achieve a new result.[1] If a highly skilled and motivated individual can complete each operation with a probability of 0.9, and the operations are independent, the probability for completing the whole sequence is given by the product of the probabilities for each of the operations; or $(0.9)^6 = 0.53$. This is just greater than one-half and much reduced from the probabilities for completing each operation. Furthermore, for a person who is just 10 percent less motivated during each operation, the probability for completing the sequence falls to $(0.81)^6 = 0.28$. Or, for a person who lacks the skill for doing one of the operations easily (so this person's probability for doing it is 0.3), the probability for completing the sequence becomes $((0.9)^5 * (0.3) = 0.18$. Thus, what first appear to be small deficiencies can have major overall effects.

One of the tasks of the effective research manager is to watch for these specific deficiencies. Often a means can be found for shoring up an otherwise capable researcher in a particular operational area. Education is almost sure to help. This might be either in a formal setting, or in an informal one created by putting the "student" in direct contact with someone who already possesses the needed skill.

If people who can execute long operational sequences are rare, it is to be expected that the productivity distribution among research workers will indicate that a few persons are far more productive than most of their colleagues. Quantitative observations confirm this.

DISTRIBUTION OF RESEARCH PRODUCTIVITY

In selecting a staff for a research organization it is crucially important to consider how productivity is distributed within groups of people. Since the total number of staff that can be hired is invariably limited, it is apparent that the organization's output per person will depend on the average productivity. If the productivity did not vary from one person to the next, careful selection of the people would not matter. Or, if productivity were narrowly distributed about some average value, random selection of the people would be nearly as effective as careful selection. Even if the distribution were bell shaped (Gaussian) with moderate width, random selection might not be disastrous because low productivity would tend to be offset by high productivity. However, as will be shown here, none of these is the case. As a result, careful selection of staff can mean the difference between a very productive organization and one that has little or no productivity.

In this context it is very important to differentiate between activity and productivity. It is relatively easy to have a high level of activity combined with a low level of the production of novel results. Thus, one of the critical tasks of the research manager is to judge the difference between activity and productivity. Unfortunately this is not easily done, and few managers do it well.

One of the first, if not the first, study of the distribution of productivity among research workers is that of Lotka.[2] His measure was the number of formal papers produced. For data, he studied both Chemical Abstracts (Decennial index, 1907–1916) for chemistry authors whose names started with *A* or *B*, and Auerbach's Historical Tables of Physics (1910) for all physics authors of note through 1900. In both cases their productivities were distributed in accordance with an inverse square law (see Figure 4.1). This is called a Pareto distribution (Montroll and Badger, 1974) after the economist who first showed that the incomes in a variety of countries are distributed in this way with exponents that lie between one and two. Figure 4.1 shows data for papers on chemistry and physics, and similar results have been obtained for mathematics.[3]

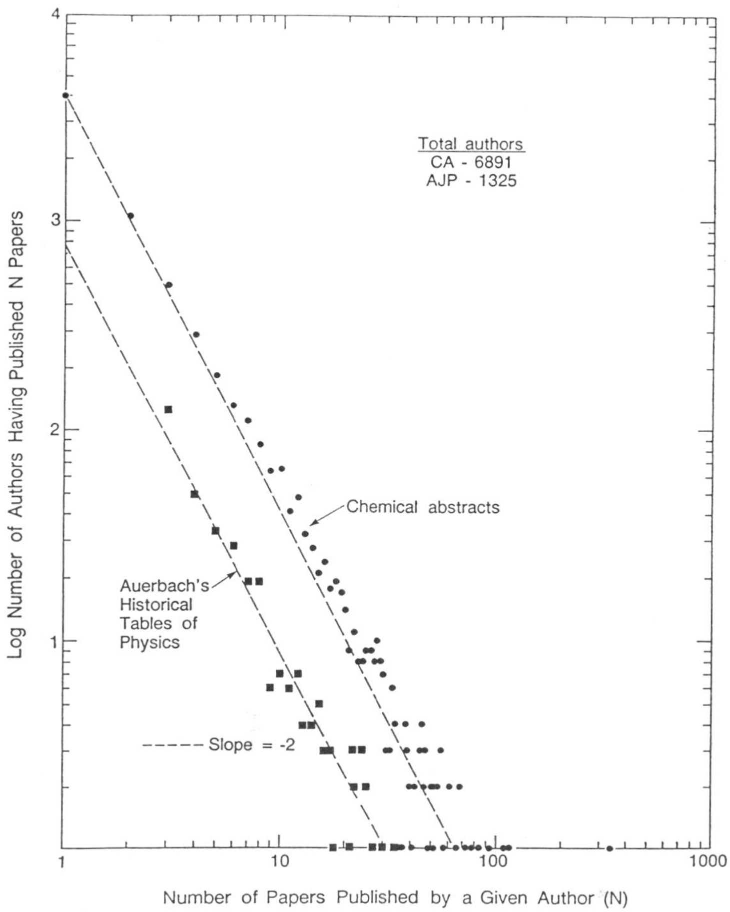

FIGURE 4.1. Publication data collected by Lotka. The straight lines correspond to Pareto distribution functions (Source: Lotka, "The Frequency Distribution of Scientific Productivity").

Shockley[4] used the log-normal distribution to describe similar data, but the inverse power law of Pareto is a better description.

Figure 4.2 shows how the Lotka data are distributed in terms of cumulative percentages. The dashed line shows what the distribution would be for uniform productivity. The actual distribution deviates markedly from it. Ten percent of the authors produced about 55 percent of the papers; and 90 percent of the papers were written by 58 percent of the authors.

An indication of the validity of using numbers of papers published as a measure of productivity is provided by the fact that there is a strong correlation between this parameter and scientific stature.[5] Various criticisms can be directed at this correlation, but as long as a high level of precision is not expected, it remains a valid indicator.

Citation studies provide some improvement over simple paper counts as a measure of creativity in that they incorporate some aspect of the quality of papers as pointed out by Price;[6] but they are not as good as patents. This is

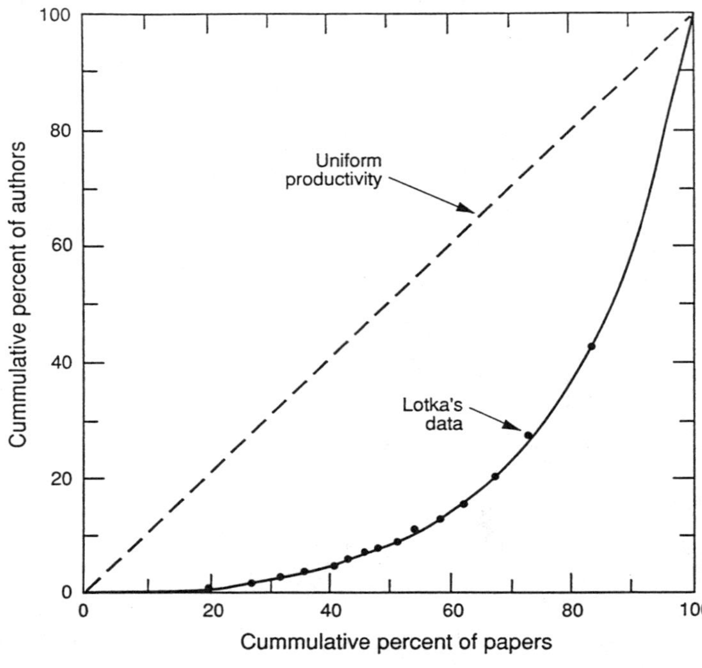

FIGURE 4.2. Lotka's data for ten years of Chemical Abstracts (Decennial index, 1907–1916); authors whose names start with A or B. Cummulative percentages are plotted (Lorenz plot). The dashed line would be expected if the productivity were uniform. The data represent nearly 7,000 authors and nearly 23,000 papers.

not to say that only the authors of patents are creative. It is a comment on the measuring tools, not necessarily on the measurees.

As a measure of technical productivity, numbers of scientific papers are improved if the papers are reviewed by the author(s) peers, but such reviews tend to be cursory because the reviewers do it in their spare time and are often preoccupied with writing their own papers.

The only literature that is seriously screened for originality and utility is the patent literature. In order to be patented an invention must be: "new, useful, and non-obvious."[7] The people who examine (screen) patent applications are professionals, and they know that their work is vulnerable to dismissal by a court if it is not done well. Therefore, the applications that survive the scrutiny and are issued patents are the best available measure of technical creativity.

For patents, a correlation similar to that for technical papers exists as indicated in Figure 4.3. This figure is based on the data given by Shockley.[8] The data follow a Pareto distribution, but the exponent is close to one instead of the two seen in Figure 4.1. This probably reflects the fact that a patent application must survive a more stringent examination for originality and utility than is required for technical manuscripts. Notice that the five individual inventors at the bottom of the figure produced nearly as many patents as the 67 people who each produced one patent (at the left-hand side).

Publications as a measure of scientific activity have had the problem that once the practitioners became aware that publications were being used as a measure, they arbitrarily increased their outputs. Not their real outputs, but the number of papers. This is easily done by writing shorter papers, writing more than one paper about essentially the same subject, writing short preliminaries followed by more complete accounts, etcetera. In other words, they deflated the currency. Thus the significance of a certain publication rate 30 years ago was different from the same rate now. Nevertheless, after appropriate adjustments have been made to-arrive at a real relative value, publication rates are a measure of productivity. This is made clear by considering the limits. At the limit of zero publication rate, there surely is no productivity. On the other hand, at the limit of infinite publication rate, it has no value as a measure of scientific or technical output. In between, it is a rough measure.

The same thing is true of patents. It is true that many patents don't have much value, but others have a great deal. Both the business and the government communities recognize that patents are a good way to encourage a supply of equity to inventors. Although patents are an imperfect representation of invention, they are the best available.

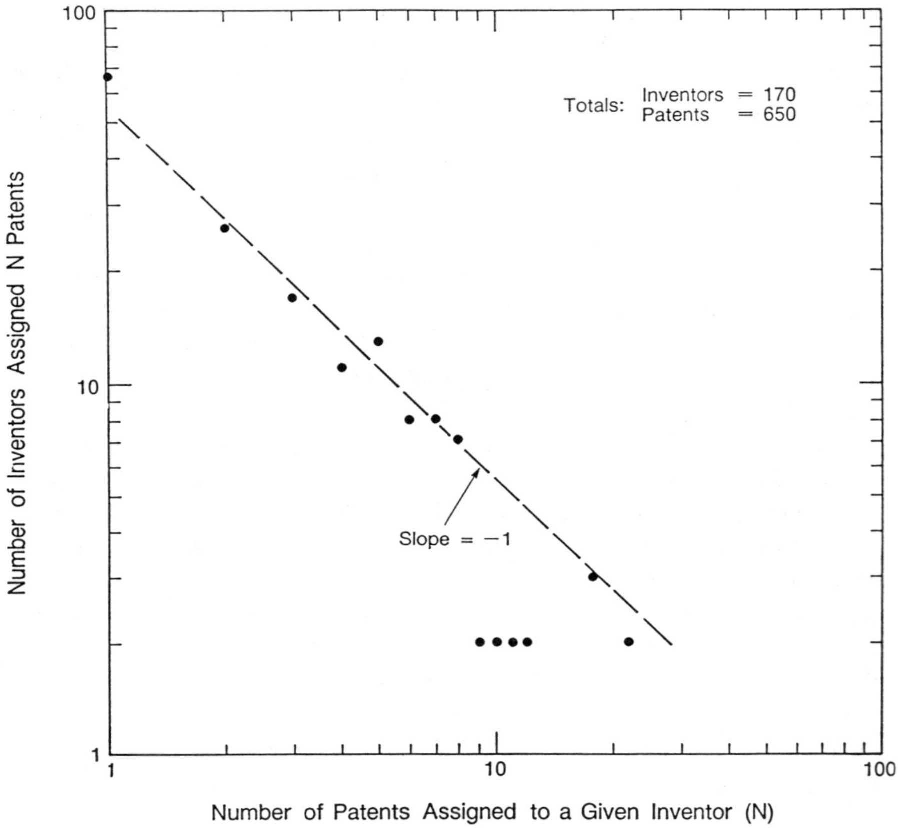

FIGURE 4.3. Showing that inventions are distributed in the same way as scientific publication. (Source: Shockley, "On the Statistics of Individual Variations of Productivity").

One of the problems of the United States is that inventivity (the number of inventions per unit cost) is going down. As research budgets go up, much of the increase is going into military activities, and other nonproductive things, plus inflation. The number of patents is not keeping up. Therefore inventivity is going down. This is more significant than the fact that the invention rate per capita in the United States is going down.

Other measures of productivity are lecture invitations, awards, and invitations to memberships in honorary societies. Unfortunately, statistical data for these measures is not readily available.

A distribution similar to that of Figure 4.1 is found for incomes. For an example, see Figure 4.4. Other things that follow the Pareto distribution

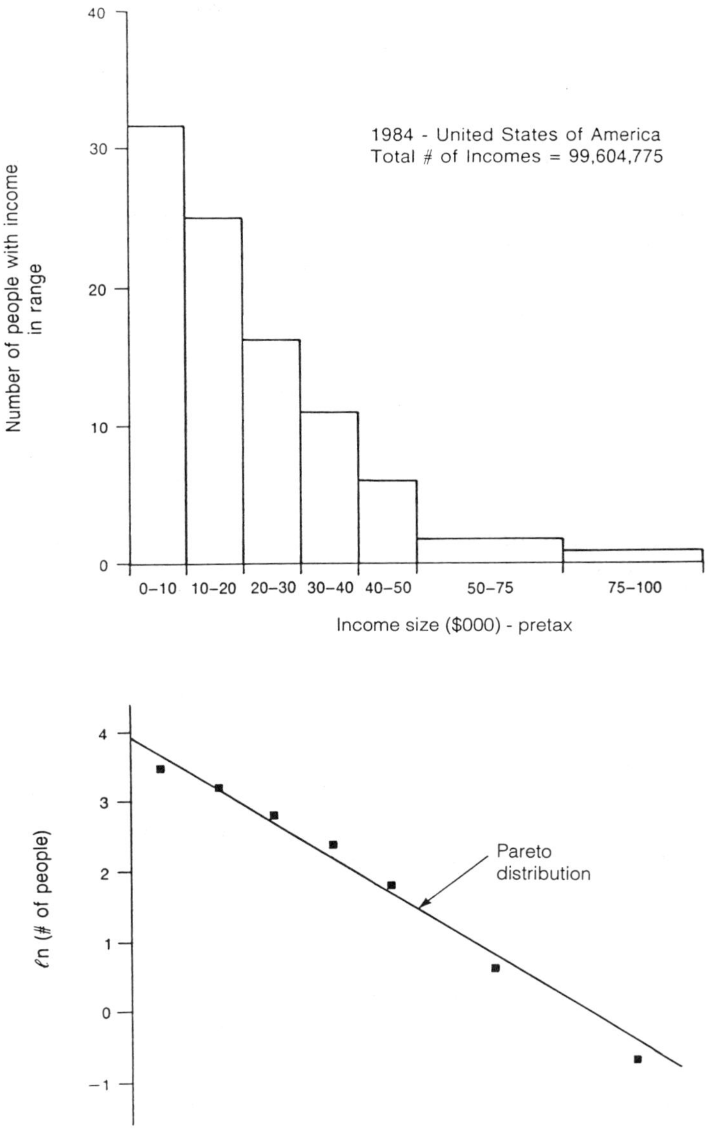

FIGURE 4.4. The distribution of income in the United States during 1984 plotted as a histogram at the top, and to show that a Pareto distribution function is followed at the bottom. Data from Microsoft Bookshelf CD-ROM disk, 1987.

function are the sizes of cities, the frequencies of words in written texts, and biological species according to genera. It also appears to describe the performance of salesmen, other types of employee performance, and the accumulation of inventories. Derivations of it have not been entirely satisfactory, tending to be tautological.[9]

However, one aspect of it is clear; it involves an equilibrium that is approached by a process in which changes are proportional to the state before the change. This may be seen in the case of authors and their papers, for example, by letting A_N represent an author who has published N papers; and B be a constant. Then a correlation line in Figure 4.1 can be written:

$$\log A_N = \log B - 2\log N$$

If this is then differentiated on both sides with respect to time, we have:

$$(1/A_N)(dA_N/dt) = -2(1/N)(dN/dt)$$

Since the two sides of this equal one another, they also can be set equal to one constant; say P. Then:

$$(dA_N/dt) = PA_N$$

so it is clear that the rate of change of A_N is proportional to A_N. In other words, the Pareto distribution indicates that authorship begets authorship, cities beget larger cities, income begets more income, and so on.

The number of authors who publish N papers does not grow indefinitely because of the minus sign on the right hand side. This constrains the growth by having the effect of decreasing the rate as N increases. The balance between the two effects results in the Pareto distribution.

Implications of the Distribution of Productivity for Research Staffing

It is apparent that the most productive researchers are manyfold more productive than the least productive. The factor can be 30, 40 times, or more. The "product" may, of course, take various forms. It may be theoretical, computational, collectional, experimental, constructional, inventional, organizational, and so on. Whatever its form, some people will produce much more of it than others. Therefore, a prime task and responsibility of the management of a research organization is to recruit the most productive people, and to effectively manage them thereafter.

The use of data on rates of scientific publications and patents in the last section was meant to be illustrative, not to indicate that the only important output is papers or patents. But the data given show how large the difference in production rates can be. It can be as much as a factor of 100. To be competitive, people whose output lies only somewhat above the average will not be adequate. They must be truly outstanding. Lack of a clear recognition of this, has led and continues to lead, to a great deal of waste—of both of money and human emotions. All too many organizations have skimpy quality and then wonder why the output is meager.

It should also be pointed out here that even an effective organization must deal with a large number of "items" in order to generate high-quality outputs. The rule-of-tens applies. Out of 10,000 of today's thoughts, about 1,000 are still worth thinking about tomorrow. Of these, about 100 are worth developing to the level of a patent application or a professional paper. About ten of these will be worth practicing as new processes or products, or will be cited by outside workers. And, with some luck, one or two of these will be a prize-winning paper, or an economically valuable invention. These numbers may seem fanciful to the reader, but I kept records at one of the industrial laboratories that I managed, and these numbers are consistent with my records.

The rule-of-tens underlies the high cost of getting research results that have real value, and the large amount of dross that is generated in the process. Small improvements can have large effects. Suppose the cost of an interesting thought is some fixed number of units (or that it takes a fixed amount of time to produce it, on average). And suppose that the rule-of-tens could be changed to a rule-of-eights. Then the previously outlined sequence would cost only 4,096 units to produce a valuable invention; a reduction of 59 percent.

IMPROVING PRODUCTIVITY

A strong statistical tendency is the regression of characteristics toward the mean. According to this, the tendency is for persons with high productivity to regress toward those with average productivity. Management's job is to counteract this statistical tendency in order to maintain above-average performance. This not an easy task given a group of people whose natural average tends to match the averages of other groups. In the long run it may not be possible, but in the short run there are many examples indicating that it can be done. How can the average productivity be improved?

The best way to improve group productivity is to get people into positions where they can do the things that they do best, and above all are not required to do things that they do badly, probably because they do not like to do them. In other words, match their motivations with the requirements of their job. Sometimes this happens naturally, or an individual does it for his, or her, self. But more often this is not the case.

Motivation is especially important in research work because it stresses a person's psyche intensely. Development work is only slightly less stressful. Edison described this by saying that making inventions is 99 percent perspiration, and 1 percent inspiration. In other words, most of the experiments (or designs) that are done by a researcher are failures. Only a few succeed. And for the kinds of achievement that Edison was seeking, less than a few succeed.

How can motivations and jobs be matched? One approach is to use questionnaires or tests of a psychological bent. I think that these are relatively ineffective and prone to distortion when expected rather than true answers are given.

Individual's have their own temperaments, which point them in certain directions early in their lives. They may have specific talents in music, art, athletics, mathematics, languages, writing, social poise, and so on. Or they may have complex profiles of skills and attitudes, which match the needs of particular types of work. As they mature, they realize certain achievements, which give them pleasure and reinforce their talents and temperaments. After a few pleasurable achievements it becomes nearly certain that an individual will seek out activities similar to those that previously yielded pleasure or a sense of achievement. Conversely, it is unlikely that an individual will seek out activities that have previously yielded failure and pain. In other words, the individual now has a "motivational pattern."

The existence of motivational patterns has been described by a well known psychologist as follows: *"A musician must make music, an artist must paint, a poet must write, if he is to be at peace with himself. What a man can be, he must be."* (Abraham Harold Maslov, *Motivation and Personality,* 1954.)

Determining the motivational pattern of an individual is a relatively simple, although somewhat time-consuming matter. Arthur Miller[10] has invented and pioneered a simple method for doing this. People are well aware of what activities have resulted in past achievements and associated pleasures during the courses of their lives. By simply asking them to list these achievements in a systematic way, a pattern can be developed. For example, they can be asked to divide their life span into five-year periods, and for each period to list what comes to their mind as the one or two outstanding achievements or experiences of that period. After a person has written down this information, it can be developed through an interview in which the various statements or anecdotes are discussed in order to clarify some of the points and to develop information about the features of the events that really excited the person.

The pattern that emerges is unique to the individual. However, it tends to have a number of universal, or repetitive, features. Its usefulness relates to the fact that it has a strong tendency to be invariant over time—both past and future. Its persistence into the future is what makes the pattern helpful in placing individuals into jobs that match their motivations. This will in turn maximize their productivities. The converse is also true. If an individual is placed in a job that requires activities that the individual is not naturally motivated to do, low productivity can be expected. Some features of the pattern that are important are as follows.

1. *What is the principal motivational thrust of the person?* Is it business? Discovery? Nurturing other people? Organizing information? Or one of many other possibilities. Whatever it is the job situation must provide an opportunity for it to be expressed. Otherwise, activities outside of the job will be used to express it.

Since the motivational pattern does not change with time, attempts to mold the individual to fit the job are very likely to fail. Even though the person's intentions are good, and he or she vigorously attempts to perform an unsuitable job, the person will drift into undesired activities, driven by his or her invariant motivational pattern.

Tragedy arises when, for some inexplicable reason, a person has pathologic motivations, and tries to do a job for which the underlying skills and motivations are missing.

2. *What abilities are used?* Is the person an observer or a doer? Does the person use language as a writer or as a speaker? Does the person derive pleasure from organizing people or data into simple patterns? Is analysis and interpretation a large component of the person's approach to problem solving? Or, is synthesis and conceptualization a more important part of the approach? Does the person tend to collect data for guidance, or use intuition instead? Is the person open to learning new techniques, or inclined to follow traditional methods?

The abilities pattern should emerge from the person's description of achievements and experiences. It should not be extracted through questions and answers because these are more prone to an interpretation of what the person thinks the "correct" answers are.

The pattern of abilities indicates the resources that the person will attempt to use in performing any job. It is important to understand and use these resources in order to maximize the person's productivity and job satisfaction.

3. *What kinds of subject matter recur?* That is, what kinds of objects, mechanisms, and subject matter is the person motivated to work with? This is important because the person will attempt to include them even if this requires that the job be distorted in order to accommodate them. Unless they can be accommodated, the job fit will be poor.

Is the person attracted to tangible or intangible subjects? Tangible means the observable and quantifiable. Intangible means perceptive and emotional. Does the person tend to be manually, or cerebrally oriented? Will the person be ingenious, or inclined to follow well-established pathways?

4. *What circumstances recur?* Does the work provide opportunity for private or for public achievement? If successful, does the work lead toward large financial rewards? Toward higher organizational position? Toward greater personal freedom? Is the person attracted to new exploratory directions? To competitive situations? To open-ended situations? To well-defined, closed-end situations?

5. *Relationship to other people.* Certain kinds of relationships emerge from each person's achievement pattern. The individual maintains this relationship through a series of experiences. For example, one person may want to work alone, another wishes to work in harness with other people, and still another wants to be the "star," the leader, or in some other role.

Whatever the way, a particular person's mode of operating with people is significant to his or her working style, since he or she will strive to maintain that mode of operation regardless of the level and type of position in which he or she might function.

The use of motivational analysis in skilled hands is almost always a valid approach to the problem of getting people into compatible jobs. Its reliability stems from the fact that individuals know themselves better than anyone else.

When I was working at Brown University I attended a seminar in the Psychology Department. The subject was the prediction of student performance in graduate schools from prior tests. The speaker discussed the correlations between such things as undergraduate course grades and standard test scores, and graduate school performance. Nothing correlated very well with one exception. This was the student's personal opinion! That is, if one simply asked the students whether they thought they were good students and could do well in graduate school, the answer was the best available indicator of their future performance.

In most working environments, the direct approach—asking someone what they are good at—doesn't generate useful answers for obvious reasons. The motivational approach is much more likely to yield honest answers because people can reasonably expect to learn something useful and perhaps impress their employers with descriptions of achievements. The experience is altogether different than that of "taking a test." It is a positive experience for all but a very few exceptions because it emphasizes positive experiences.

In dealing with the people of a creative organization, it is my experience that indirect methods of trying to enhance performance are better than direct ones. First, because they "pull" instead of "push"; creative people will expend much effort in pushing back, if pushed. Second, because they are not "Mickey Mouse" exercises, and therefore cannot be frustrated by the application of "reverse Mickey Mouse." Third, while indirect methods may generate chores, they do not compromise independence in any serious way. This contrasts with methods in which quotas, objectives, and goals are assigned. These latter lead to a great variety of game playing, which is a waste of management time and is destructive of employee morale, as well as management credibility.

There is always a conflict between the collective needs of an organization and the individual needs of its members. The balance in this conflict can never be all one way or the other. However, the optimum position of the balance varies. In an army, for example, it necessarily lies on the side of the organization. This is also the case in production oriented organizations. On the other hand, in a creative organization, the creative members must come first or the organization will cease to be creative.

I have known people, including managers, who seem to have operated with the idea that productivity is proportional to the amount of activity. This may be true of back scratching but little else that I can think of. Activities need to be coordinated so an output is generated. Thus it is the coordination that is important, not the activity. Without coordination there is no true output. The problem of enhancing productivity becomes that of teaching improved activity coordination.

Improving activity coordination is such a complex process that no recipe for it exists. It can be learned, however. One approach is through the use of informal "annual reports." This is discussed more in Chapter 12.

RECRUITING

A result of the points already discussed in this chapter is that of all the functions that a research manager has, the most crucial is recruiting. This is clear if the implications of the Pareto distribution of output is combined with the implications of motivational analysis. The latter indicates that motivation should not be expected to change significantly for people whose patterns have emerged. The former indicates that output varies enormously from one individual to another; so much so that it is extremely unlikely that a low-output person is convertible into a high-output one, even if motivation were changeable, which it is not. It must be concluded that performance is largely determined at the time of recruitment.

Very large companies have the advantage of placing new recruits "in training" (probation) in their research organizations. Those not best suited to the work can then be moved into their operations. This practice is advantageous to all—at least in principle.

While the research performances of individuals can be marginally improved through experience, practice, and education, they cannot be drastically changed. This again puts a premium on the recruitment process. The best way to optimize recruitment that I know of is to make decisions about candidates collectively. That is, hold a meeting of all the persons who have interviewed the candidate, and elicit opinions of the candidate's characteristics from each interviewer in the presence of the others. Then, after some discussion, make a decision—by voting if necessary (usually a consensus will develop one way or the other without a vote). The collective wisdom of the group is more likely to be correct than the wisdom of any one individual. This is why democracy works better than monarchy.

Strong candidates for recruitment will effervesce with enthusiasm for their past and present activities. This may not be overt but will be unmistakable, even in quietly spoken individuals. They will be "full of" their subject. They will be anxious to carry it forward.

Weak candidates will be reticent. They be "willing to do anything that is requested of them." They will lack conviction about their past work. If asked to describe their past accomplishments, or the type of work that they like best to do, they will have trouble identifying it clearly.

Having made a positive decision about a candidate, one way to improve the odds that the candidate will accept an offer of employment is to make it immediately. That is, before the candidate has left the premises. This makes the candidate think that he or she is especially desired by the organization—a powerful aphrodisiac. Since superior candidates are always in great demand, whatever the organization can do to tip the odds in its favor should be done.

Remember that employment decisions are almost always critical for a candidate so small forces can have very large effects. The candidate is in a state of unstable equilibrium such that small influences will move him or her in one direction or another.

In selecting candidates for employment, it should be remembered that school records are incomplete predictors of success. A good school record indicates competence, but not an ability to invent or do other creative work. Track records are far more important. Also, it is always essential that successive recruits bring either different, or improved, skills into the organization. This is true whether it is a new organization working at building its strength, or a mature organization that must maintain its strength. The organization must constantly try to raise its standards. This is the only way it can increase or maintain its productivity.

If an organization is successful in progressively increasing its strength, and therefore its productivity, it will not be a comfortable place in which to work. Internal competition will continually increase. The earlier hires will feel pressure from the recent hires, and so on. But comfort does not yield productivity.

The people in a successful research organization will be idiosyncratic (originals), and rebellious (dissatisfied with the *status quo*). Not a pleasant prospect for a management that wants its people to be quiet, problemfree, and obsequious. But a pleasant prospect indeed for a management that wants fire, excitement, and inventivity.

CHARACTERISTICS OF THE PEOPLE OF ACADEMIA AND INDUSTRIA

Recruits for careers in research and invention pass through academic institutions with very few exceptions. If they go into academia, no substantial transitional event occurs. But, if they go into industria, they encounter a substantial change of values. Motivations, styles, personalities, and other factors become quite different.

Underlying the value differences is the instability in industria that is induced by the time value of money. There is no *status quo*. Turnover is essential because inventory that is not being liquidated generates interest charges. Real estate, equipment, or raw materials that are not being used effectively are not neutral, they are liabilities because they are assets that are not producing returns. Proceeds that could be obtained by liquidating liabilities could be invested to obtain assets. The result is that the time value of money (the time derivative of the value) is a parameter that is necessarily ever present in the milieu of industria. By contrast, this parameter is largely ignored in academia.

Fortunately, instability has more than one direction. It can lead to decay, but it can also lead to growth. If industrial assets are well-managed, they can rapidly generate cash that can be applied to research and invention. If these are successful in creating valuable assets, more cash leading to more research may become available. Thus the reward system in industria is often more direct than in academia, and is another reason for the different characteristics of their inhabitants.

A good description of several of the differences between academic and industrial people has been given by Furnham and Pendleton.[11] A paraphrased version of their listing is shown in Table 4.1.

LEADERS OF THE ENTERPRISE

A group of research people will have little effect if the leadership structure to which they report is faulty. Therefore, some of the conditions for it to be favorable are listed here, and then discussed. This section is based on conversations with George Kazan (see Chapter 6 for a brief biography).

1. The company's chairman, or the real power leader, must be a research proponent, and must be absolutely sold on the idea of expansion–diversification through research.

TABLE 4.1. Comparison of the characteristics of people who work in academia and industria (Furnham and Pendleton, "The Gulf that Divides Us").

	Responses	
Characteristic	Academia	Industria
General aims	Insight and knowledge	Action and operations
Speed	Low urgency	High urgency
Output type desired	Elegant and critical	Applicable and clear
Input data base	Direct empirical	Second-hand empirical
Complexity level	Frequently complex	Frequently simple
Uncertainty treatment	Statistical	Personal
Presentation mode	Written documents	Face to face
Personal facade	Irrelevant (often shabby)	Crucial (fashionable)
Persuasion mode	Empirical data	Rhetoric
Cost–Benefit Analysis	Irrelevant	Crucial
Personality Type	Introvert	Extrovert

2. The company must be prosperous. The research effort must not be viewed as an incubus by leaders of the company's operating groups.
3. Major people must be found to lead the research. If it is a new laboratory, a new director must be found; either internally, or from the outside. To really do the job, the new director must be a national leader, and nationally recognized, in the field of choice.
4. The programmatic theme, or themes, should be original for this sponsor, and should originate with the appointed leaders. It is important that the leaders themselves know what they want to do. That they don't ask the sponsoring company, "what should we do?". That's not the right kind of leader. And, they must be people who have good batting averages in projecting future events.
5. The new director must attract for subordinates technical leaders who having been operating at the state-of-the-art level. These are the people who will be in charge of the subdivisions of the laboratory. And that must be true for every new theme that is concocted.
6. The hiring program must be budgeted and contracted for at least five years, and a critical number of people must be hired to do the research work. Small trial efforts are a waste of effort (money).
7. The research themes must be carefully selected. This is the most critical of the choices to be made. It means that the younger people should select the themes. Very few 50–60 year-old people will be as fresh as 30–40 year-old people. The older ones will tend to be defensive of decisions they made earlier on.

The chairman, whether new, or an existing employee, must be an unswerving backer. He or she must not vacillate, or waiver. All the decision makers, such as the executive committee, or the chairman's coterie, must be in agreement. The chairman must be willing to sign-off, that is contract for, at least five years worth of capital requirements, operating costs, buildings, equipment, instruments, salaries, and materials. Participation in the careers of the new people, either by transfer or by hire, is also important. The chairman must agree that these people are good selections, and know the people and their histories well; and be a colleague of the new people, be collegial. Chairpersons should interview them, probably more than once, and they should, of course, be a participant in the hiring negotiations. They must understand the hiring agreements, and stand behind them in name and actions. They must approve of and continue to support the pace of staffing and its budgetary consequences. No short-range problems should be allowed to interrupt. The new theme, the new idea, is the number one priority.

Chairpersons should recognize that the development of the technology after the research phase, and the development of markets is very expensive and time consuming; much more so than the research itself. They should understand this, clearly knowing what they are getting into.

If the chairpersons are not as good as their signatures, and the business literature indicates that many are not, the environment is wrong. Effective research cannot be done without a contract. That is, a firm understanding between the corporation (represented by its business leaders) and the research organization (represented by its technical leaders). When the contract is broken, for whatever reason, the plan begins to fall apart. The plan becomes penalized. It becomes less than it was. It is not a case of going from 100 to zero. It is not a wipe out. But, suddenly it becomes a 90, then a 75, a 30—and its gone.

No one should attempt to lead a research organization unless the chairperson of the company in which it exists can be trusted to support it through good times and bad. It is the chairperson's responsibility to provide for the future of the company, and the future is what research is all about. If they are not willing to provide a support contract, the environment is wrong. And the support should come from them personally, not in some diffusive second-hand way.

People who can do research and lead it effectively are rare. They should not be asked to risk their careers for a few coins. This may be idealistic, but the quality of research and the research profession is not likely to suffer from idealism. Many current problems stem, in fact, from a neglect of idealism. Research leaders should not be put into positions where they are working for people who will behave toward them in a fickle manner. Dedication toward the task must be long-term and genuine.

Many business people treat research as if it were a long shot at a betting track. They expect it to produce instant wealth, or nothing. This attitude is almost sure to produce failure. As explained in earlier chapters, the probability is small that wealth-producing inventions will come from any particular research activity, and if it does it will probably not come quickly. Therefore, if efforts are not to be wasted, they must be properly managed in a style suited to optimizing the path to success.

The company's prosperity is very important. Cash-flow must be good, making it difficult for the old guard to complain about the expense.

The research leader, either from an internal transfer or a new hire, must personally be at some science frontier within the theme that has been selected. Personal involvement at the very edge of some important facet of the theme is a must. The leader should have a solid academic background plus some industrial laboratory experience at one of the great invention centers. He or she should be young in spirit—attitudinally between 35 and 40. Virtually boundless energy, unquenchable curiosity, tenacious memory, and a genius for organizing time are all needed. The leader should have wide knowledge of peers, a good reputation in the eyes of peers, be well-known in a field of specialization, have prolifically published papers of quality, know peers personally, and be able to articulate projects to them persuasively. The leader should be ready to spend the rest of his or her career in the post. These specifications are somewhat idealistic, but there are people who satisfy them.

Staff members should be hired from excellent schools with the intent of providing lifelong careers to them. Some of them should be encouraged to do science. All should be allowed to write, publish, and seek peer recognition. The more they can be made to feel like individual entrepreneurial professionals, the harder they will work and the more creative they will be. However, regarding publications, no company can afford to publish before it establishes its patent position. So there is often a delay.

Since the output of scientific activity is papers, people who do not write and publish papers are not doing science, inventors to whom patents are not issued are not doing inventing, developers who do not produce prototypes are doing something else, just as are commercial people who do not sell products. These activities vary from one part of an organization to another. Some people are not supposed to be doing science, so naturally they don't necessarily produce papers.

For people who are supposed to be developing hardware, some of the questions are: Does the machine appear on time? Does it work? Is it competitive? Is a new process put together in a timely fashion? Is it superior to older processes? Is it economical?

For people doing commercial development work, it soon becomes obvious whether they're selling a product. Their effectiveness is easy to measure.

It is a consequence of the educational system in the United States that an unrealistic fraction of the people working in industrial research laboratories wish to consider themselves to be scientists working at doing science. It seems to be very important to their morale. They maintain the illusion, even when they must know that it is only partially true, or only true in a small way. But it's important for them to think of themselves as scientists. They talk about "doing science." They say, "I am doing science." Whether it's false, it is part of the persona that they have crafted for themselves. They try hard to maintain academic ties through recruiting trips and the use of academic consultants.

The practice of putting academia on a pedestal is probably one of the reasons why other countries appear to be more effective in transforming inventions into businesses than the United States has been in recent years, and than Great Britain has been for many years. In Japan, as well as some European countries, the research required for the Ph.D. degree, or its equivalent can be (and often is) done in industry or some place other than a university. This has had a powerful effect on attitudes toward nonuniversity research work. Much of the best work is done in industrial laboratories and research institutes rather than in academic departments.

A most important characteristic of the research boss is to know what he or she wants to do from the very start. The matter of program should be discussed very early. By the time the successful candidate for the job of research boss is hired, he or she should have been asked to write down a plan for the proposed research effort. The candidate should have been able to state what themes are intended for pursuit and be able to discuss theme subtopics. This clearly puts the burden of deciding what an industrial company should do on the candidate. He or she must then demonstrate an ability to generate contemporary, viable, pertinent programs. Judgments about long-range topics will emerge from the process. It will become clear how well the candidate knows the literature, and who the players are. It should demonstrate his or her commercial viability through some of the thoughts about materials, chemicals, devices, and components that are discussed. It is important to look for schemes, or elaborate systems, among the various thoughts. Grand schemes such as the chemical transformation of crude oil to nylon polymers, to yarns, carpets, and dozens of other fabricated products might emerge.

Candidates should *be* at the frontiers of their technical specialties. Being part of the frontier virtually guarantees their competence. They must know the content of their areas of concentration. Otherwise they can not be at the frontier. They will also know the other first-rate players because they will have been in competition with them. If they are not personally at the frontier they will not be able to lead the work, no matter what their administrative skills may be. They must be dedicated to making significant advances in the chosen area. Devotion to, and passion for, technical accomplishment are essential if their performance is to be more than mediocre. Candidates must exercise good judgment in the face of myriad unknowns. This requires an intuitive grasp of the subject possessed by only those at the frontier.

Managers of the groups that report to the research boss should be picked for their technical competence in the topics that have been selected for emphasis. They should be qualified to replace the current boss in a few years; a condition that is difficult to meet. Through their professional reputations they should be qualified to debate their peers and to judge them. Incumbent peers should participate in the selection of recruits. They should be people who have already demonstrated their leadership.

Technical leaders must not only have high technical credentials, they must also have high tolerance levels for unusual personalities—as described by Shaw in the heading of this chapter. Others are unlikely to be inventors or discoverers. The most successful persons will be those who are driven to solve problems of some kind that others have not been able to solve.[12] More reasonable persons will be satisfied with the *status quo*. Not so for the problem solver.

NOTES

1. E. W. Montroll and W.W. Badger, *Introduction to Quantitative Aspects of Social Phenomena* (New York: Gordon and Breach, 1974) pp. 110–120.
2. A. J. Lotka "The Frequency Distribution of Scientific Productivity," *Jour. Washington Acad. Sci.*, Vol. 16 (1926) p. 317.
3. H. T. Davis, *The Theory of Econometrics* (Bloomington, Indiana: The Principia Press, 1941).
4. W. Shockley, "On the Statistics of Individual Variations of Productivity," *Research Laboratories, Proc. IRE*, Vol. 45 (1957) p. 279.
5. *Ibid.*
6. D. J. de Solla Price, *Little Science, Big Science... and Beyond* (New York: Columbia University Press, 1986).
7. S. Lubar, "New, Useful, and Nonobvious," *Invention & Technology*, Vol. 6 (1990) p. 8.
8. Shockley, "On the Statistics of Individual Variations."
9. G. Yule, "A Mathematical Theory of Evolution, Based on the Conclusions of Dr J C Willis, F.R.S," *Philosophical Transactions of the Royal Society*, Vol. 213B (1924) p. 21. For other derivations see: H. A. Simon, *Models of Man* (New York: John Wiley & Sons, 1957) Chapter 9; J. Steindl, *Random Processes and the Growth of Firms* (New York: Hafner Publishing Company, 1965); and D. Sahal, "A Formulation of the Pareto Distribution," *Environment and Planning*, Vol. 10A, (1978) p. 1363; W. W. Badger, "An Entropy-Utility Model for the Size Distribution of Income," in *Mathematical Models as a Tool for the Social Sciences*, ed. B. J. West (New York: Gordon & Breach, 1980); H. A. 0. Wold and P. Whittle, "A Model Explaining the Pareto Distribution of Wealth," *Econometrica* (October 1957) p. 591.
10. A. F. Miller and R. T. Mattson, *The Truth about You* (Berkeley, California: Ten Speed Press, 1989).
11. A. Furnham and D. Pendleton, "The Gulf that Divides Us," *New Scientist* (November 18, 1989) p. 66.
12. R. S. Root-Bernstein, "Who Discovers and Invents," *Research-Technology Management*, Vol. 32 (1989) p. 43.

5

HOW SHOULD THE WORK BE APPROACHED?

Langmuir emphasized the futility of making formal plans in an effort to obtain new ideas. He stressed the importance of developing a receptive mind which could accept a new idea engendered by a fortunate accident, an unexpected occurrence, or some other set of circumstances beyond the immediate control of the individual.

—Vincent J. Schaefer in "Can We Do It Better?", Bull. American Meteorological Soc., Feb. 1968.

Effectiveness in research work is very different from efficiency. In the creation of inventions effectiveness is essential, whereas efficiency is not.

Imagine a research organization administered by the most modern methods, including centralized data processing and communications. Populate it with inexperienced people chosen to minimize salary costs consistent with good, but not outstanding, quality. Make sure that written communications pass from each individual to every other individual regarding all aspects of planning, budgeting, executing, and evaluating. Take pride in low personnel turnover arising from either raids from the outside, or purges from the inside. Keep careful control of marginal costs through meticulous bookkeeping. Have frequent drives to cut marginal costs. During training courses, performance evaluations, and when allocating rewards, emphasize promotion to the management ranks as the only personal goal worth pursuing.

The imaginary organization just described may well be quite efficient, and well administrated. However, its effectiveness in producing inventions, or other creative results, would be close to nil. Therefore, in spite of its superficial appearance of great efficiency, it is not well managed because it would not fulfil its purpose.

Efficiency is of prime importance in a production organization. A more or less continuous stream of product (either hardware or software) is needed in this case. For a research organization, on the other hand, the output can be quite discontinuous. That is, it can consist of irregular spikes. It should be considered successful if the spikes contain significant results.

Teamwork is prized in production organizations, and rightly so. But discovery is a private matter. It occurs within a single mind, so teamlike activities may encourage it, but cannot cause it. This may be recognized by managers but is rarely acted on.

More than once I have heard a manager say: "the role of a subordinate is to act so as to make his boss look good." When I hear this, I make a mental note: "Ouch! This guy is a turkey, more interested in himself than in getting the job done." Why? Because this statement is inverted. It should read: "the role of a boss is to make his subordinates look good; if they look good he will look good." The boss does not do the work. He or she should guide it, and coordinate it, but someone must actually *do* it. If the organization is to be effective, the doers must *do*, and as a consequence look good. This will push the whole organization forward, including the boss.

In the first mode of the previous paragraph, the boss may "look good" in the short run, but subordinates' time will be consumed doing it, so the long-term results of the organization will suffer. The boss who wants to take "looking good" from his subordinates, instead of giving it to them, is obviously not interested in the welfare of the organization but in personal advancement.

A closely related syndrome is that known as "I'm OK, you're not OK." It manifests itself as, "I'm busy with very important matters, what is the question you're bothering me with?" Or, "Me boss—you peon—why aren't you working instead of bringing your problems to me?" Or other forms of arrogance. It is frustrating indeed for workers to attempt to discuss their problems with a boss only to have them ignored while the boss unloads his problems onto them. Here is an appropriate quotation:

> "A friend of mine characterizes leaders simply like this: 'Leaders don't inflict pain, they bear pain.' "
>
> —Max Depree, Chairman of the Board,
> Herman Miller Inc., *New York Times,*
> Section 3, p.2, September 10, 1989.

A deeper result of the attitudes previously described is that they make real communications between boss and subordinate impossible. This, in turn, means that working together toward common goals does not occur. Instead, antagonism develops with each side trying to "get something" from the other.

One of the most important differences between high- and low-output research organizations is the balance between the attitude of the organization toward itself as contrasted with its attitude toward its employees. One extreme exists when the organization considers itself to always be right and expects its employees to frequently genuflect. The other extreme occurs when the employees ignore the legitimate needs of the organization and are considerate only of themselves. Both attitudes are destructive in their extreme forms. However, both need to be present, so one of the tasks of management is to balance them.

Unfortunately, in recent years there has been a steady drift toward modes of financing research activities that exacerbates the problem outlined. Research is treated as if it were a commodity! Its activities are divided into projects that sell for fixed prices. And its people are divided into pools of expertise and are assigned to particular projects for relatively short terms of time. There is much talk of "matrix management," which denigrates skills and talents of the people attempting to work under it. Marketing, rather than technical skills, become prized. The individual worker increasingly becomes a pawn in a game of increasing fantasy. When it is recalled that the basic unit of creative work is an individual person, it is clear that this mode of behavior does not lead toward strong organizations with high productivities.

Nucleation Versus Modification Versus Revolution

A most general characteristic of research is that it is an agent of change. Change comes in small, medium, and large sizes, of course. Small changes, or modifications, are easily effected. They can be carried out nearly homogeneously by simply announcing them to an organization. Large changes are qualitatively different. In revolutions, attempts are made to impose change homogeneously (that is, uniformly upon everyone in the organization) by dictatorial fiat. Very mixed results almost always ensue. Revolutions can be achieved much more effectively in an inhomogeneous mode, as is well-known to physical chemists when they observe phase transformations. That is, they can be achieved through nucleation and growth.

Nucleation of change means that a relatively small part of an organization is given a charter that is distinctly different from the rest. Its operating rules may also be distinctly different. There must also be a driving force behind it that gives it some stability and some opportunity for growth. Otherwise, it will simply shrink out of existence. After it has grown sufficiently large, it will continue to grow spontaneously and eventually consume much, or all, of the previous organization.

As organizations grow larger, their toleration for distinctly different nuclei grows ever smaller, so the nucleation and growth approach to change is used less and less. No organization can be expected to be comforted by the existence of a colony of foreigners in its midst who look different, speak a strange technical jargon, and may have good fortune on their side.

Take the DuPont company as an example. This is convenient because of the excellent history of their R&D operations that was published recently.[1] Formal research was nucleated in the company in 1902 by the establishment of their Eastern Laboratory. Initially the work of this laboratory strengthened their explosives business considerably. One result was the growth of research in the company and the establishment of the DuPont Experimental Station. But, by the mid-1920s, the research organization was being criticized within the company for not producing new products. So Charles Stine proposed the nucleation of a new department to be devoted to "fundamental research." This was eventually done, Carothers and his colleagues were hired, and nylon and neoprene were born. Within ten years, however, the department had been decimated, although the reputation of research had been restored. Twenty-five years passed. Dissatisfaction with research had reappeared. By this time, the company had grown so large that another "nucleation experiment" could not obtain the approval of the company's Executive Committee. A result has been continued frustration. And a lack of dramatic new results.

CENTRALIZATION VERSUS DECENTRALIZATION

The product lines of small, new firms, and a few large, old ones, tend to be monolithic. They are tightly focused onto a group of closely related products. It is obvious that such firms should have their research efforts centralized. For small firms because this is the only way that a critical mass of research workers can be built up. For large firms it is not so obvious, but it is difficult to take serious exception to this mode of organization when the company's product line is homogeneous.

As a company grows it tends to become diversified. It tends to create a set of operating divisions each with a rather distinctly different product line. Should it now have centralized, or decentralized research activities? The answer, of course, is both. However, the charter of the centralized group should be quite different from those of the decentralized groups. In practice, they often all have essentially the same charter with an understanding that the time scales are expected to be different. That is, the centralized group is expected to have longer range projects, but all projects are expected to hew closely to the existing products of the company. This is a recipe for disappointment and frustration!

The most general difficulty is that under the arrangement just described no one is seriously searching for future opportunities outside of the existing product lines, which are inevitably maturing, or becoming obsolete. The next most serious difficulty is that the responsibility for maintaining a high level of technical expertise, not to mention leadership, in each of the operating divisions becomes abrogated. The operations managers take the position that it is not necessary because the experts, if needed, are available in the centralized organization. In addition, it becomes virtually impossible to recruit and sustain outstanding people in the operating organizations if the people in the central organization are considered to be the experts.

It may be concluded that if an operating division is large enough to support a viable research group with a few percent of its revenues, it should. In addition, it should contribute a percentage to a central corporate research group.

Doing Versus Studying

Doing was previously mentioned. Another way of stating the case is that actually making things in a research organization is more important than describing things or manipulating information. By this I mean making specimens, apparatus, chemicals, microorganisms, devices, software, and so on. All great laboratories have had notable shop facilities because this is how new entities get synthesized. In the British tradition Davy was renowned for his apparatus, Faraday followed him, as did the Cavendish laboratory of Thompson, and of Rutherford. In the United States one can cite the laboratories of Edison, General Electric, DuPont, Bell Telephone, Eastman, Lawrence, and Land.

The shop services of all of these have been outstanding: stockrooms; machining facilities for dealing with metals, wood, polymers, and ceramics; capabilities for analyzing chemical compositions, structures both small and large; fabrication groups for making glassware, instruments, electronic circuits, and optical systems; and graphic arts departments. Many of these services have become so specialized that they are now done by independent service companies, but they still must be substantial items in the budgets of organizations that hope to be inventive. Reducing costs by cutting them back is false economy.

On Delegation

Delegation is not a natural activity. The natural impulse is to do it yourself. But, in an organization of almost any size, this is not practicable. I once worked for a man who thought that his capacity for work was limitless. His position put him in charge of several hundred people. Given his flawed view of his capacity, plus his view that no one else could really do things right, the result was disastrous. Long queues of decisions were left waiting. Fear was prevalent, especially because he couldn't remember all of the previous decisions he had made since there were so many of them.

It is relatively simple to assign tasks to subordinates. The hard part is giving them the authority that they need to execute the tasks. The latter involves giving something up, so it is a particularly difficult thing for an insecure personality to do. Also, it is often not clear *a priori* what is being given up. That is, the superior's conception of the amount of authority that is needed to do the task may be an underestimate. Thus the subordinate may create some unexpected problem. Now the superior becomes the reluctant owner of a dilemma. The superior may undercut and thereby alienate the subordinate, or support the subordinate's position, thereby antagonizing someone else. The resolution of such a dilemma is a simple, but clear-cut, test of good management. The reaction of a poor manager will be to chastise or by-pass the subordinate. The good manager will support the subordinate, and deal with antagonizing others separately.

The manager who fails to delegate adequate authority along with responsibility makes it impossible for subordinates to operate in a manner that is consistent with responsible decision making (that is, decisions for which they accept responsibility for the outcome, whether it is favorable or otherwise). Therefore, they will simply stop making substantive decisions, and all of these will flow to the top of the organization where they will tend to stagnate and perhaps fester. This might be an acceptable mode of operation if improved decisions resulted from it, but this is unlikely. It is a very old adage that the best decisions tend to be those that are made at the very lowest level in the organization where adequate knowledge of the situation is available. If they are moved to higher levels, less complete knowledge is available, so the odds of poor decisions increase rapidly, no matter how much intrinsic wisdom exists.

DOES MANAGEMENT MAKE A DIFFERENCE?

My answer to this question is clear. However, members of the "less is best" school are probably skeptical. On the other hand, research people who have worked in more than one organization can often produce anecdotal evidence (or opinions) indicating the effects of good, and less good, management. Excellent management is usually discounted by researchers, of course, in favor of individual research talent, but this is a near-sighted view.

Objective evidence is provided by the patent literature together with business data. With all of its faults, the patent literature is systematically studied by professionals for originality and utility. Well-managed research organizations have higher inventivities; that is, more patents per unit of expenditure. In other words they generate originality more effectively.

A study of the patent literature[2] has shown that for a set of 145 companies in the period 1976–80 the R&D expenditures per patent varied by a factor of 125 from about 10^5 to 1.25×10^7 dollars per patent. The set consisted of those companies that were granted at least one U.S. patent per month (on average) for each year of the five-year period. That is, companies that were serious about making and protecting inventions. For government research organizations where patentable inventions are often not a goal, the ratio can reach 10^8 dollars per patent, or more. But, on average, government inventivity is similar to the low values of the largest companies. Thus, for the decade of the 1980s, the average number of patents issued to government activities was about 1,108 per year, while the 1990 research expenditures were 16.3 billion dollars,[3] so the approximate cost per patent during that period was 1.47×10^7 dollars per patent.

Part of the large variation in the cost per patent is because of different management goals at different companies. Part is also because of differing methods for reporting "research" expenditures. Another part of the variation is related to the size of the organization (as indicated by the size of its budget). Figure 5.1 indicates this effect. The regression line shows that the number of patents obtained for a given cost declines with the amount of the cost (that is, in most cases, with the size of the company).

The size effect can be interpreted in various ways, but it is consistent with the idea that decentralization is good for research productivity. The opposite is true for development projects, so once again, the evidence is that one style of management does not fit all activities.

If the variations of inventivity are caused simply by a lack of definition in financial reporting procedures, then these need to be reformed. Otherwise, companies will not understand what they are buying when they make research expenditures. On the other hand, if the variations are caused by the social dynamics within large organizations, then improvements in inventivity might be obtained by using decentralized organizational modes for research staffs.

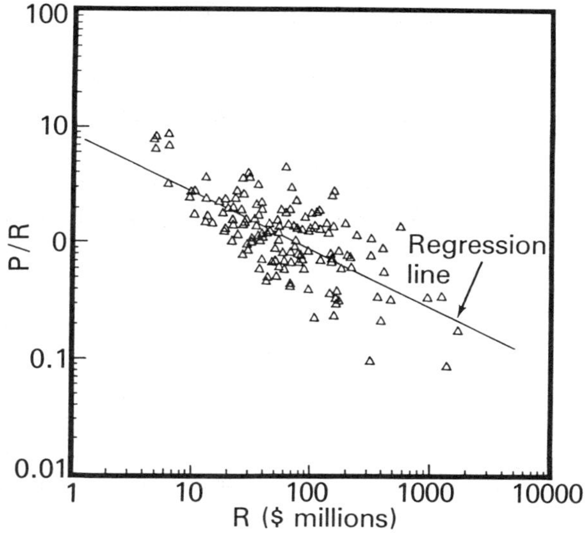

FIGURE 5.1. Correlation of P/R (# of patents per year/millions of $ expended) with R (millions of $ expended on research). Data for 145 companies averaged over 1976–80 are shown (Source: Patent Intelligence and Technology Report (Alexandria, Virginia: IFI/Plenum Data Company, 1981). The correlation coefficient is −0.70.

Protection of Intellectual Property

The principal method, and probably the most effective one, for protecting intellectual property is patents. It dates back to the Constitution of the United States of America, which says:

> *Congress shall have the power to promote the Progress of Science and useful Arts, by securing for limited Times to Authors and Inventors the exclusive Right to their respective Writings and Discoveries.*

Why do pharmaceutical companies tend to enjoy high profitabilities? Why is research a more stable enterprise in the chemical industry than in most others? Why are companies that manufacture a great invention like the laser rarely profitable? There are many reasons, of course, but one of the most critical ones is how an industry interacts with the patent, trademark, and copyright systems.

Some people believe that inventions should belong to no one. That they should be completely nonexclusive. Such people tend not to be inventors. The latter often believe that inventions should be owned by them exclusively—and forever. The patent system provides a compromise between these two extremes by granting temporary exclusivity to inventors. It has served the chemical and pharmaceutical industries well.

It is instructive to compare the experience of invention in the field of modern electronics. Since many of the early inventions associated with integrated electronics and lasers were made at the Bell Telephone Laboratories, the patents have been available to everyone. The result has been open price competition among the producers of products, exacerbated by excessive numbers of participants in the marketplaces. This has destroyed profitability for most of the producers, and has made bankruptcy among them commonplace. It has also allowed foreign competition to maraud freely, particularly from the Far East.

In contrast to the situation with modern electronics, the classical era was more orderly (relatively speaking, there certainly was ample tumult). The telegraph, telephone, dynamo, electric lamp, vacuum tube, and other patents were given at least minimal respect.

Open competition is a false ideal. The nearest independent farmer will verify this. The very purpose of open competition is to drive business margins toward zero. But even an ideal business cannot be viable with zero margin. New businesses start small. To achieve minimal efficiencies they must grow at least until they reach a size that is stable. They can only grow if their revenues exceed their costs, and this requires substantial margins. The quantitative aspects of this are discussed elsewhere,[4] and will not be repeated here. The point that I am trying to make is that the patent system has great value. It should be strengthened rather than being subjected to denigration as it has been in some quarters in recent times. The same is true of the trademark and copyright systems.

A key feature of the patent system is that the franchise it provides is temporary. Thus it does not protect against competition indefinitely, only for a set period of time. This allows new businesses to get started and old businesses to recover their technical and commercial development costs.

Patent protection helps increase inventivity by increasing the incentive for making inventions and offsetting the costs with improved investment returns. New procedures for protecting intellectual property are badly needed as commercial activity shifts from manufacturing durable objects to producing various forms of organized data (information products).

An absence of patent protection is a major disincentive for innovations because secrecy alone will not protect them now that fast and sensitive methods of chemical and physical analysis are widespread. These allow almost any product to be "reverse engineered," and produced at relatively low cost since the large initial investments in research and development do not need to be amortized as part of the manufacturing cost.

Large companies tend to have lower inventivities than smaller ones as pointed out in the preface of this book and by J. Herbert Hollomon in his study of the patent system some years ago.[5] At that time, in spite of misconceptions to the contrary, independent inventors and inventors in small companies received 60 percent of the patents issued in the United States, 5 percent were issued to inventors in government and nonprofit organizations, and the remaining 35 percent to inventors in large companies.

It might be argued that although the number was smaller, the quality of the patents issued to large companies was larger. The evidence is not clear because some outstanding inventions were patented by both groups: the large companies and the small companies, or individuals. What is clear is that large companies are not highly efficient producers of inventions. It costs them as much as two orders of magnitude more to produce an invention than it costs a small operation. Why?

One quite real possibility is that they have a bias against hiring true inventors, as contrasted with people more interested in science than in invention. This bias is societal. Inventors do not enjoy the same level of status as do scientists, in spite of the fact that inventing is a more difficult art than scientific investigation. Large companies tend to reflect the biases of society more than do small companies.

Another important reason is that invention is not perceived to be as important to the businesses of large companies as they are to small companies—especially in the short term. For many small companies an invention may be the entire basis of their franchise. But for large companies, other factors such as finance, marketing, and manufacturing efficiency are important parts of their franchise.

Protecting inventions is made difficult by the all-too-common academic bias against it. This is expressed as the imperative to publish accounts of the results of technical work. The imperative is held to be more important than the retention of exclusive knowledge by a sponsor whom has made an investment that has supported the inventor and his work. The bias is propagated through the students who pass through the academic system without being exposed to other views of the issue. It is often wrapped in the cloak of "academic freedom," forgetting that freedom is not the same as license.

Both sides of the issue of protection for inventions are represented by self-serving interests. Therefore, the ethics are not clear-cut. They seldom are for this or any other issue. One side wishes to advertise itself through publishing technical results that may be acclaimed by others. The other side wishes to exploit the invention through commercial development.

NOTES

1. D. A. Hounshell and J. K. Smith, *Science and Corporate Strategy, DuPont R&D, 1902–1980* (New York: Cambridge University Press, 1988).
2. J. J. Gilman and A. Siczek, "Optimization of Inventivity," *Research Management* (July–August 1985) p.29.
3. Report to Congressional Requestors, *Technology Transfer*, GAO/RCED-91-80, U.S. General Accounting Office, Gaithersburg, MD (1991).
4. S. Ramo, *The Management of Innovative Technological Companies* (New York: John Wiley & Sons, 1980).
5. J. H. Hollomon, "The U.S. Patent System," *Scientific American*, Vol. 216, (1967) p. 19.

6

Strategies of Many Varieties

It is amazing how few oil people really understand that you only find oil and gas when you drill wells. You may think you're finding it when you're drawing maps and studying logs, but you have to drill.

—John A. Masters, Oil explorer.

Fifteen strategies associated with research programs will be described here. This will not exhaust the possibilities but will give some sense of how variable they are. The specific examples that are used are based on the experiences of George Kazan. Someone else would have a different, albeit related set. The strategies given here are drawn from events that occurred at the Allied Chemical Corporation (now Allied-Signal), the Chemstrand Company (now Monsanto), and the DuPont Company. Thus they relate to the chemical industry, mostly toward the end of its golden age while it was a growth industry that could do no wrong. Similar events could be drawn from other industries, but these have the advantage of being described by someone who participated in them first hand, and shed much sweat during some of them. Kazan, trained as a chemical engineer, was an active participant in most of the examples given at virtually all levels of management from bench research through process engineering, large plant design and construction, commercial development, general manager, research director, and corporate vice-president.

The style here is different than in other chapters. It is in the form of an interview. This took place in Berkeley, California, on June 7, 1989.

To help put the following events into context, the Allied Chemical

Corporation needs to be characterized. Originally known as the Allied Chemical & Dye Corporation, the company was formed by merging five large chemical companies in about 1920. The result was the largest chemical company in the United States. The five were: General Chemical (acids and inorganics); Solvay Process (alkalis and chlorine); National Aniline & Dye (dyes and organics); Semet Solvay (coal, coke, and coal tar derivatives); and The Barrett Company (building products). After the merger, the component companies maintained their autonomies, and competed with one another. Initially, this kept the company unusually prosperous and innovative but not expansive. Based on the company's own ingenuity, a Textile Fibers Division was launched in 1958 and enjoyed prosperity. To ensure its feedstocks, the company bought another new division in 1961: Union Texas Natural Gas Company. For 50 years Texas Natural Gas was a prominent producer of fertilizers and other agricultural chemicals. Until the late 60s, the company was proud of its internal technology, which was generated by its engineering, process development, and research staffs. At that time, centralization began to change the character of the company. It was no longer the largest U.S. chemical company principally because it was slow in entering the synthetic fibers arena.

The first form of strategy to be discussed, *Type I*, is the "grand scheme." It consists of choosing a broad area of technology that appears to be fertile for development. This requires a prophet and the prophet's conviction that it will be possible to obtain development through invention, investment, and organization. It is difficult to define all this succinctly, but it should be clarified by the discussion that follows.

GILMAN: We are going to talk about the history of polyethylene innovations (waxes and pipes), at the Allied Chemical Corporation (now Allied-Signal) as an example of how a grand scheme develops.

KAZAN: Yes. The key thing about these polymers is that they are specialty polyethylenes. In a 10-cents-a-pound world for standard polyethylene, they sold for 50-cents-a-pound.

One of them has very high molecular weight (over 1–3 million), and one of them has very low molecular weight (1,600–1,900). At the time, no one else had made branched, low-density, waxy, low-molecular-weight polyethylenes. None that I know of at least. The waxes were first made (about 1948) in a high pressure reactor (3 kbar) in Morristown, NJ, from cylinder ethylene. It was done in a brand new Central Research Laboratory, who's charter was to put Allied into new businesses. This was a new concept for them: operating a corporate R&D

Laboratory, all by itself, with no dependence on any operating group for laboratory work, for people, for programs, or the development of programs.

The first program I know of at this place was initiated by a decision to make polyamide fibers independently of other companies. To have an all Allied polyamide based upon technology they developed for themselves. They were, of course, following DuPont. This was post–World War II, and they were following work started at DuPont in the 1930's. This work had been exploited during the war, and amidst the craze for women's nylon hosiery just before the war (1940–41). Polyamides were used for making parachutes, shrouds, special binding formulations for straps, ropes, etcetera, taking advantage of the very high strength of the polyamides.

At the same time the Allied people decided to develop polyamides, they decided they wanted to go into polyethylenes. These had come to this country because DuPont and Imperial Chemical Industries (ICI) had a cross licensing agreement for patents. It is critical to this story that we know that this agreement allowed DuPont to be the first manufacturer of polyethylenes in this country, from a discovery made by Imperial Chemical Industries in England in the mid-30s. Under the agreement, DuPont got the process and the product formulations, and enough engineering to design a plant. Wartime needs for polyethylenes were supplied by ICI. After the war it was obvious that polyethylenes were useful, and DuPont started to build polyethylene plants, letting the world know that it was doing this.

Allied, putting ideas into its brand-new Central Research Laboratory, hired a polymer chemist from the Brooklyn Polytechnic Institute who was a student of the renowned Herman Mark, who consulted for Allied. The student was told to put Allied in the polyethylene business. This became a grand scheme, dramatizing the R&D process for chemical systems that involved: (1) getting raw materials, (2) processing them, (3) making finished materials, (4) finding uses for the finished materials, and (5) finding customers.

Allied liked grand schemes in which oil and gas and minerals in the ground are converted into products that customers use. Allied could control all the raw materials, all the things that were done to the raw materials, and all the postprocessing that made them into polymeric materials.

ICI's patents were read at Allied. DuPont had announced to the world that it was building at its Sabine River Works in Orange, Texas, a large

ICI-type polyethylene plant to operate at 50,000 psi (3 kbars). The research people had no apparatus that could operate at pressures like that. They had a "bomb" that could operate at 0.5 kbars (8,000 psi), and it was used to start an experimental program. It used a free radical generator (isopropanol and a peroxide) as the reaction medium. I have forgotten which peroxide was used—an easily available one, vintage late 1940s. They got reaction products when they pumped the bomb full of high-purity ethylene. The products were waxes with molecular weights of about 1,700 (about 100 repeating CH_2 units). They kept synthesizing low-polymer waxes without being able to bring them to much higher molecular weights. They tried varying the peroxides, varying the ratios, varying the alcohol, varying the hydrocarbon, varying the purity of the ethylene, and other things that I don't know about. All variations gave waxes.

The transition of these waxes into an invention, and later commercial entities, was made by Frank Porter, who was the research boss. Porter said to the group one day, "What's wrong with waxes? Why don't you develop a business based upon them?"

At that time they didn't visualize anything but homopolymer waxes. They didn't visualize copolymers or terpolymers. They also didn't visualize oxidized systems where the chemical reactivity of the polymer was changed by deliberately introducing oxygen. Oxygen in the polymerization process itself was a no-no. The oxygen treatment to make a polymer polar was a postsynthesis step.

Porter's question played a key role. The polyethylene group wouldn't have asked itself that question. It was after a commodity business, the emerging ICI–DuPont high polymers. Another big producer of high polymer at that time was Union Carbide.

The Allied researchers found that the waxes were not brittle. They were tough, and they could be chemically modified. They had mechanical properties that did not limit them to waxlike applications. The first big use was in lined dairy cartons. And there were other uses that a group was organized to find. Bob Rosenbaum was the leader of this group for the rest of his career—about 30 years. He developed uses in inks, in rubber compounding, in polishes for cars and shoes, and in extrusion modifiers. The waxes were an effective releasing agent for molded parts, good antifriction agents, and also were used as carriers for high polymers. The markets developed at a steady rate.

It was easy to see by 1952 that a plant had to be built. They chose to build it in Buffalo, New York, because they had ethylene there. The ethylene came from coke-oven by-product processing. There was a big coke plant run by Semat-Solvay in Buffalo, which put out a gas that included a sizeable amount of ethylene. Lummus Engineering built a 20-million-pound ethylene plant at Buffalo to separate the ethylene from the rest of the constituents of the coal–tar gas. So ethylene was available for polymerization, and the project was assigned to the Semat-Solvay Division.

Thus the Central Research Laboratory broke away from the original idea of operating independently of other people. At about the same time it was decided that the development of the nylon fiber operation would be done by the people who synthesized dyes and pigments. As dye makers they knew the textile markets, so it was concluded that nylon should go to them. This is what actually happened; I'm not endorsing the logic.

The plant was built around a very large pressurized reactor (0.5 kbar). It was on River Road in Buffalo about 1,000 yards from a big steel plant. They started making wax, but they didn't understand the catalyst system well. In preparing the system for the initiation of polymerization they would inadvertently allow air to get in. At least I think that air was the villain. Gradually, they got much better at keeping it out. In the mean time, they had very bad explosions. They blew every glass pane out of the steel mill, every one. River Road is a major highway going from downtown Buffalo to Niagara Falls, and they blew the windows out of cars. I don't think they ever hurt any people, but the town of Tonawanda outside Buffalo began to complain.

The explosion problem was solved by more care in the synthesis, and the product started to sell. By 1956 they were selling 4–5 million pounds a year at 50 cents a pound. The marketing operation was centered in New York City and Morristown. Their first chore was to become knowledgeable about the milk business, the problems of various milk containers, and the problems with paraffin as a liner for a paper carton. And how to print on cartons, what to do about reducing the number of leaky cartons, what to do about sealing the closures, and so on.

Rosenbaum, and his people at Semet-Solvay, kept developing new uses. Porter, the director of the Central Research Laboratory kept funding new applications projects as well. By 1957 they were selling up to 12 million pounds. At a time when polyethylene began to drop rapidly in price, they were still selling waxes for 50 cents a pound. This kept on until they had used up the 20 million pounds of ethylene from the Lummus plant. There was no more synthesis gas in Buffalo so they went to Orange, Texas. The new plant was expanded probably four or five times until the business reached 60 to 70 million pounds a year. The mill cost of the polymer (not the complete cost) was about 8–9 cents per pound, and the product was very profitable. For a division whose business was buying coal, and making coke, this was splendid new business.

GILMAN: We should switch to the high-density story now. This begins in 1954–55 with the inventions by Phillips Petroleum for novel polyethylenes.

KAZAN: These were polymers with molecular weights of about 60 or 70 thousand. They were essentially linear, but not completely. They had a density of around 0.95 instead of 0.91, and they had good mechanical properties. They were strong, but they were notch sensitive. If they were scratched, cracks would propagate through them. This problem was solved by copolymerization techniques. Phillips was trying to sell licenses to various companies, including Allied.

At the same time, in Germany, at the Max Planck Institute, Karl Ziegler invented another linear high-density polyethylene that was rheologically different, and whose chemical structure was different. Although they were similar, the two products were distinct. Ziegler sold licenses everywhere and tried to sell Allied. Everybody—by everybody I mean Koppers, DuPont, Allied, and the oil companies—had polyethylene fever.

The Phillips disclosures had shown that their polyethylenes were made on catalyst supports like silica, or silica-alumina. The supports were impregnated with CrO_3, and then activated. Allied turned down both Ziegler and Phillips; they liked to do things by themselves. Dr. Joris, a physical chemist, was assigned to develop a process based on silica-alumina as a support, and magnesium chromate and dichromate as a catalyst activated under a certain ritual that was developed. When the first polymers were made, in high-pressure reactors, they were rheological absurdities. They were very high molecular weight, and they were almost completely linear. The silica-alumina supports were cylinders; roughly 10 mm. in diameter and 12-13 mm high. These were impregnated with the catalyst, and then suspended in hexane, or pentane, or isooctane. For good measure some aluminum alkyl was added. So the catalyst system was a sort of hybrid of the Ziegler technology, which was based on aluminum alkyls, and the Phillips technology, which was based on chromium oxide catalysts (not dichromates). The Allied people put all these things together and got new polymers.

The new polymer looked like a mushroom. At the end of the reaction it was cylindrical with the same proportions as the catalyst pellet. The pellet material was incorporated uniformly throughout the polymer product. The biggest product pellet I ever saw was about 10 cm in diameter by 15 cm high. There they sat—these great big, fuzzy cylindrical hunks of polymer—in the high-pressure bombs. And nobody knew what to do with them!

Enter Frank Porter again. This time he said "What's wrong with high molecular weight?" It had fantastic impact properties, so it could make wonderful hammers. The researchers could think of other things to make from it, but how would you form it? It was too stiff to press into a die, and you couldn't push it through an ordinary extruder; it wouldn't move. But Porter got the idea that it was going to be great for plastic pipe.

Other people had started making experimental polyethylene pipe. DuPont had started, for example. Their product's name was Alathon. Porter was determined to make pipe from the new Allied polymer. He suspended the high molecular weight species in wax (20 percent wax and 80 percent high molecular weight polymer). He then extruded this blend at about 400°C, and 5,000 psi. It moved very slowly out the end of the extruder. So he built a die with a mandrel and started making pipe. He became really enthusiastic when the material didn't fatigue under cyclic loading.

Strips of this material could be bent back and forth all day long. They wouldn't break. They were rubberlike. That encouraged Porter because pipes flex with alternations in internal pressure so they need fatigue resistance. He got the idea that the stuff would never fatigue. Also, it didn't have a lot of the pinholes and other flaws of DuPont's Alathon. But it always extruded slowly, so Porter devised a continuous pressure molding scheme in which the extruder was connected to a big manifold, and a lot of smaller branches coming from it with pipe dies at their ends.

Typically, a master extruder (100 pounds an hour) was connected to a manifold with 20 small extruders coming from it. Low and behold pipe came out of these things at 2 to 3 inches per hour. It was like a giant cow with 20 nipples. It was very slow and very unstable, and the polymer was not reproducible.

The first pipe samples had fantastic properties. Allied had been selling bitumen to the Orangeburg Pipe Company for making tarpaper pipe for sewers, house drains, gutters, etcetera. They were the first customer. They put a full-page add in the *Wall Street Journal* and got ready to start making pipe with dies, manifolds, the small extruders, and the windup devices all fabricated at Allied's machine shop in Morristown.

GILMAN: How did they stabilize the process? You said it was unstable.

KAZAN: The extrusion process never did get completely stable. For another problem, ultraviolet instability, they added carbon black. But the very high-molecular-weight polymer wasn't always the same. Sometimes it would extrude as intended, and sometimes it would not.

I became the project manager. I had to make catalyst, take it up to Buffalo, and put it into the reactors. These were designed so the heat of polymerization could be dissipated. I developed a technique for putting the catalyst system into giant rotating cylinders; they were 10 feet in diameter and 60–70 feet long. They were rotated by a pinion and a giant ring gear set on huge bearings like a kiln. The ethylene was introduced in batches through a sleeve at the center of the head.

By 1957 we had five of these cylinders making polyethylene. We sold it to every major extruder in the country.

Trying to keep the catalyst batches uniform was difficult. Some wouldn't work at all. I worked on that problem too. I was the overseer for making all the tools and equipment, and I sold the product to customers. We had to install all of this equipment in their plants. A man named Hoppler and I did that.

We had to find the customers. The Department of Commerce had put out a standard for tubular goods (say garden hoses), and the tubular goods people had a trade association. Through it, I sold the idea that we could make a plastic pipe with certain properties, and I got a Department of Commerce standard established. It said that certain of the properties had merit. Then I got a National Bureau of Standards certification that the properties were meritorious. But there was no labor union that would install this pipe.

We had a customer named Carlon in Cleveland. The owner knew people in the union, and they helped us try to win them over to the idea of using plastic pipe. This was before rigid polyvinylchloride pipe was available. There was PVC garden hose at the time, but no PVC pipe as we know it today. After a year of trying, we got the union to go along with us.

By 1960, we were selling about 8 million pounds of pipe a year, and we had 5 big customers—everybody was happy. We had two specialties: the pipe, which sold for a high price; and the raw pipe compound, which could be extruded into pipe.

Overall, this was a case in which the key people did everything along the way. There was nothing that they wouldn't try to do. To sell the material we eventually had to install equipment in an extruder's plant, and then teach him how to use it to form shapes.

Allied Chemical began to dream of very-high-molecular-weight polyethylenes, with very high volume. In 1962, this led to the purchase Union Texas Natural Gas company to feed raw materials into Allied's plastics program. It became extremely profitable, especially during the oil crisis of the 70s. The operating profits in 1973 approached one billion dollars.

So the polyethylene work became a grand scheme with many inventions along the way. The wax reaction system, wax applications, the catalyst system for the very-high-molecular-weight polymer were all inventions. The systems for compounding the high-molecular-weight material, for building the extrusion apparatus (I always thought of it as a continuous compression moulding technique) was also innovative. And it was very, very profitable.

GILMAN: It's interesting that, on the one hand, the specialties were the high-profit items, but the revenues were dominated by the commodities.

KAZAN: Yes. I think that pretax income from the waxes in 1960 approached 15 or 16 million dollars, and the pipe about 1 million dollars. Thus the specialties earned much more than commodity medium-molecular-weight material. Its earnings were probably 7–8 million dollars.

The most successful result of the whole program was the acquisition of Union Texas Natural Gas. Its total profits from 1962–90 have been many billions of dollars.

A *Type II* strategy, that I'll call the "major theme" strategy, is one that I was personally involved with—namely, fluorine chemicals. There were some fluorine chemicals at Allied before the time of this description (roughly the period from 1940 to 1970). It's an example, not so much of a research idea, but of a thematic strategy that led down many paths. If it could be isolated in all its aspects, I would say that it represented a gain in net income for Allied of about 25 percent. To get 10 things that are big gainers, you also have a lot of losers that you don't count. But the net effect of exploiting that single element, fluorine, was a 25 percent gain in net income. It was won over a 30-year period—it didn't come suddenly.

When I became the technical vice-president in 1968, I found four groups making fluorine chemicals. They were making inorganic fluorides, elemental fluorine, and organic fluorine compounds. There was a group at the National Aniline Division concerned with textile applications, taking advantage of the electronegativity of fluorine. There was a group at the General Chemical Division concerned with refrigerants and propellants. There was a group in the Central Research Laboratory concerned with more esoteric things like membranes. And there was a fourth group for Defense Department applications. The total number of technical people working on fluorine in 1968 was about 50. The work was decentralized according to the market interests of the sponsoring hosts.

It started in 1936, when a chemist named Herb Pearson was hired by the General Chemical Division's research laboratory. He was the first fluorine chemist hired from Duke University, and he was a student of Lucius Bigelow. Herb was a promoter. He had done his dissertation on the fluorination of benzene with elemental fluorine. It was a dissertation that told how he did it, what he got, how he identified what he got, and how to improve the selectivity of the reactions. Herb was an organic fluorine chemist. He stayed with this topic throughout his career. So Pearson had part of the idea.

In 1939 the president of the General Chemical Division was an engine person. He had worked in Detroit during World War I on the liberty aircraft engine, and brought this background to General Chemical. His name was H. O. C. Ingraham. He was very dissatisfied with the productivity of the research laboratory at Laurel Hill, Long Island, along the Gowanus Canal. I worked there for two years. I was hired by Ingraham's new entry, John Donleavy, who was an organic chemist from Yale. Donleavy contacted Pearson and they started talking about the future of fluorine chemicals. At the time, Allied Chemical made small amounts of aqueous hydrofluoric acid, a rat poison from sodium fluoride, and some sodium bifluoride—a chemical used in detergents. The amounts were small, and the business not very profitable. Pearson and Donleavy, talking to me about it, undertook to get into fluorine chemicals in a more highly forced way.

They started by slowly hiring more of Bigelow's students. By the 1950's, 14 years after Herb Pearson was hired, they had hired about 10 of them. They were organic fluorinators. One of them was a very good friend of mine. I got to know him well, and what he did at Duke. He became the first inventor of this group by inventing a process for the addition of HF to acetylene, making an ethylidene fluoride. That was the beginning of a major theme.

The first initiating event was the adoption of the theme. The second was the invention of a patentable method for adding HF to acetylene. The third event was the end of World War II, when General Chemical investigators wandered around in Germany, picking up alien-property custodian-packages. They found a package in a mine in Germany, which contained cylinders of fluorine, fluorine trifluoride, chlorine trifluoride, bromine trifluoride, and technical information. Maybe there were some other interhalogen compounds. They were in a wet mine, they were of steel, they were in very bad condition, and they were very dangerous. The profitable results that I will mention couldn't have been achieved without these three initiating events.

We also had very sad accidents. If there was a silica inclusion in the steel, the fluorine would dissolve the siliceous material, sometimes creating a hole. Then, if the cylinder had any liquid fluorine in it, which would be under 3,000–4,000 psi pressure, it moved around like a rocket. We lost a man that way. Fluorine at 3,000 psi coming out of a small hole, would pick up the whole cylinder.

The first major products of the work were refrigerants and propellants. Kettering and people working with him at the General Motors laboratory had searched for safe refrigerants to be used in household refrigerators and small air conditioners. Midgley, who was the inventor of tetraethyl lead, led the search. Together with a fellow on his staff named Albert Henney, and mostly on paper, they decided that dichloro-difluoromethane would be a safe refrigerant. At the time, the president of General Motors was Pierre S. DuPont because his company owned 50 percent of GM. So the two companies interacted. DuPont made some dichloro-difluoromethane, and the properties confirmed the forecast. Freon™ 12, as it was called, was stable, nontoxic, and nonflammable. As was trichloro-monofluoromethane: Freon™ 11. Eventually a whole family of chlorofluorocarbons was developed.[1]

To exploit the invention the two companies formed a third one called Kinetic Chemicals. DuPont owned 51 percent, GM 49 percent. It was quite profitable.

Allied, through D. S. Young, discovered and patented the method for adding HF to acetylene. The next step was to convert the acetylenic 2-carbon compound to a 1-carbon compound. On the ethylidene fluoride, there is an easily replaced hydrogen, so they replaced it with chlorine. Then they thermally cracked the material in-the presence of chlorine. This yielded dichloro-difluoromethane (Freon™ 12) plus carbon tetrachloride. The yields were poor, but they kept getting better and better. Using this process, they built a plant to make Freon™ 12. Later, in 1959, DuPont's patents expired, and Allied built a new plant using DuPont's process instead of its own. But they were already in the game.

Encouraged by the success of the fluorocarbons, Allied marketed many new products containing fluorine. They started with ultrahigh-purity anhydrous-HF, which was used for the alkylation of isobutane and butylene to make isooctane. This high-purity material was specially prepared. It is what I spent World War II doing, freeing the HF of sulphur dioxide, silicon tetrafluoride, and water.

The next chemicals that were made were boron trifluoride, fluorine, sulphur hexafluoride, and other interhalogen compounds. Another one was cobalt trifluoride, which is a fluorinating agent. A big customer for that was the AEC plant at Oak Ridge. Then, in the period between 1948 and 1951 they developed a 5,000 ampere generator to make elemental fluorine. It worked very well. It ran 30 days (700 hrs.) before the anodes failed. Because we could make fluorine well, we bid on a plant to make uranium hexafluoride. At that time (1955–56) we had developed a process to make UF_4 from UO_2 with HF. We reduced precursors to make UO_2 From this, we made UF_4 and then added more fluorine to get to UF_6. That's how we got into the uranium hexafluoride business, and that's why the government built a plant that Allied operated in Metropolis, Illinois.

Roughly at the same time, in about 1958, Allied started to make tetrafluor-ethylene. Then they started to polymerize this to make Teflon™—they called it halon TFE. It was also a very successful product. The process never developed the bad problems that DuPont had with explosions. Also we made soil repellents for textiles and carpeting; they were fluoridated organic compounds. Another branch of fluorine chemistry that was developed under the program was polytrifluor-ethylene. This was used in a small scale during World War II in the Manhattan Project. We used it to make films and moulding compounds that were exceptionally stable and had exceptional barrier properties. The basis for much of this work was the fluorine generator, which continued to be improved. Problems with the anodes were alleviated by inventions, and were aided by licensing some technology from ICI.

Much of the elemental fluorine went to make UF_6. The rest went to compounds such as the high-voltage dielectric made for General Electric: sulfur hexafluoride. Large amounts of this very stable compound went into transformers. The UF_6 plant in Metropolis ran for a long time. In 1973 it was still operating very profitably. One of the inventions that made this plant successful was a novel fluid-bed reactor for the reduction, hydrofluoridation, and fluorination steps.

Another product is boron trifluoride, which is still made in Claymont, Delaware. It is made from boric acid and HF.

Most of these products are still made and sold by Allied-Signal. They were retained by the parent company through various mergers.

Fluorine-based products is an example of the *Type II* major theme strategy. Eleven or 12 major products came from Donleavy and Pearson, and their successors. It's also an example of a strategy that the chairman of the company wasn't aware of. There were entrepreneurs. One was Herb Pearson. He was in at the very inception of the use of propellants, such as those used for hair spray or shaving cream. He was one of the first people who developed chlorofluorocarbons for propellants. Another was Hans Newmark. He was the entrepreneur for elemental fluorine.

Every one of the projects had patented inventions. The original process for making Freon™ 12 had patented inventions in each one of the four chemical steps. The fluorine cell had patentable inventions. So did the sulphur hexafluoride. We were the first people to make sulphur hexafluoride. We never had any real competition. Our price was limited by what General Electric could use as an alternate dielectric.

The time span was very long, and at the beginning the markets were all secondary. The markets for propellants were secondary, as were those for sulfur hexafluoride. Interhalogen compounds like chlorine trifluoride never did sell very well.

Let me list some other examples of major themes. Allied nylon, for example. Monsanto's organophosphates plant for plant hormones and other plant chemicals was another. DuPont's urea or nitrogen-based pesticides, and certainly DuPont's dye stuffs, starting much earlier, became a very large one. The largest of all in the chemical industry was DuPont's fibers program.

Type III strategy is the exploitation of a star researcher (most of this section is based on information from Hounshell and Smith[2]). In 1925, a man named Charles Stine at the DuPont company began to sell the idea that DuPont had become too decentralized and that the decentralized laboratories (which were large even in 1925) worked on their existing products. They worked on cellophane, rayon, explosives, and dyes, and they were very good at what they did, but they never generated things that were substantially different from the existing products. I'm going to discuss later the acquisition strategy of the DuPont Company, but it was also a problem because it placed demands on the use of the available talent.

Stine finally got executive committee approval to start a central research laboratory. They called this laboratory "Purity Hall." One of the first people they hired was Wallace Carothers. He was hired in about 1928. He was one of a half-dozen prominent young organic chemists of the late-20s in the United States. Carothers was hired to work on organic polymers. He was very difficult to hire; the hiring negotiations took a long time. He had lots of doubts, and they promised him the moon. They promised him a substantial budget, laboratories, and supporting staff. They promised him the right to publish, they promised him the right to spend time in coordination with the university laboratories where he had worked before. And from his work on polymers came two major DuPont inventions. His organization was like a Carothers Laboratory; although it was actually the central research section of the Experimental Station. From Carothers' work came nylon 6-6, other nylons, and the idea that polyamides were very useful materials. The second thing that came from his laboratory was the discovery of neoprene. So neoprene and nylon were the big things, with nylon much more financially successfully than neoprene. Both required major technical developments.

There are other examples of exploiting a star such as David Baltimore and the Welcone Foundation at MIT. Genentech is another enterprise gathered around superstars. Allied Chemical's Materials Research Center was another one.

Type IV initiating strategy is licensing a major invention. This is a way in which you can provide an operating arena for inventive people. For example, nylon was licensed to Monsanto (originally Chemstrand) by DuPont under duress. This became a place for Monsanto inventors to invent, and they did much for it. Allied-Signal did the same thing when they licensed polyester yarns, and fine denier nylon yarns. These became an area where inventive people worked.

Type V strategy is an order to put the host company into a certain business by fiat. Allied nylon is a good example. The Board of Directors decided in 1943 to go into the nylon business, and they built a Central Research Laboratory in Morristown, NJ, around nylon. That was the origin of the Morristown laboratory. It was done by fiat. There was nobody there to do it. There was no building, there were no people.

They were allowed to hire a few people from within the company, one of whom became the entrepreneur. That was Glen Nesty. They were designated by the chairman and president. Everybody was involved in it. The charge from the Board of Directors was given to Silsby, to Porter, to Nesty. The inventors assigned to the task were vital.

Their first marketing target, set by fiat, was fiber yarns. This necessarily developed major interests. They had to develop one great big chain of processes stretching from benzene and propylene to caprolactam, then to polycaprolactam, and to fiber yarns. It took at least ten years to do it, maybe more. In every step of the way, from making cumene, to making cumene hydroperoxide, to making phenol, then to making caprolactam by three steps, and to making polycaprolactam, inventive people were given the motivation to work, the arena in which to work, and subjects to invent into. They were given something to start with that was substantially improvable.

GILMAN: That's how Allied got into the phenol business?

KAZAN: Yes. Allied got into the phenol business because of nylon. Allied got into the caprolactam business because of nylon. Allied became a big merchant of caprolactam, as well as a big merchant of phenol. In this case of starting by fiat, the chairman is involved, the time span is shorter, and the markets are better defined than for most other strategies.

Another strategy, *Type VI*, is when management decides to buy someone else's major invention and develop it. The best example here is when the Scientific Design Company had developed the new Mid-Century Process for making terephthalic acid. It was an incompletely developed process, and they sold it to Amoco (Standard Oil of Indiana at that time). Then Amoco did a major development job on it, and put it into production. By 1985 they were producing something like 3 billion pounds of acid per year. It was an incredible bonanza. The oxidation process for making terephthalic acid, and the improvement of the purity of the terephthalic acid for the manufacture of polyester fiber became the foundation of Amoco's petrochemical business.

Initiating strategy *Type VII* is war. Wars breed experimental activity. War creates laboratory arenas for inventors to work in. The chairman has to be agreeable. Frequently a wartime affair does not have an entrepreneur in it. The inventors have a field day. Times are short during wars. An example would be the Grace Nuclear Division. Exactly how it started, I don't know. But they got into the fuels business, and into the fuels reprocessing business, because there was a war, and because there was a Manhattan Project. Nuclear power plants for naval ships as large as cruisers still use Grace fuels. They found the nucleus of this activity in Oak Ridge, bought into it, and stayed there. Some of the war-initiated effects are larger.

Synthetic dyes in this country, for example, became a huge business. German dyes were cut off in 1915. In 1914 and part of 1915 German submarines landed in the United States carrying dyes. A favorite port was Baltimore. That was stopped when the submarine warfare started. The stoppage stimulated DuPont to get into the synthetic dye business, which provided it with expertise in organic chemistry, which in turn led to its great successes in polymer chemistry and other branches of industrial organic chemistry.

Another thing World War I did was furnish the United States with diphenylamine, a dye intermediate. It was the first antiknock compound that Kettering found, and it was a wartime development. Still another war baby is Teflon™. It was a curiosity before the war. Allied fluorine, SF_6, and HF, were war babies. So wars nucleate lots of inventing effects.

Next consider *Type VIII* initiating strategy. This type has to do with the precipitous, almost panicky rush into a science that's in the air. In 1948–50 polyethylene was in the air. At first, polyethylenes were called polythenes, and were made by DuPont in the mid-30's and licensed to others by DuPont. During the war, radar polyethylenes came from the British. After the war, under their cross-licensing technique, DuPont built two ICI plants. Zeigler started to publish on polyethylenes from the Max Planck Institute in Karlsruhe, and Phillips developed other polyethylenes. Polyethylene, although the markets weren't clearly defined, caught the imagination of the chemical, and petrochemical worlds. They succumbed to a fad. In this case, initially the chairman knows nothing about it. There's no entrepreneur. There's an inventor. We have already talked about him for the Allied chemical ethylene project. The time span is much shorter. For the sale of Phillips polyethylene, and Zeigler polyethylene, it was about five years.

Type IX, the next mode of-initiation, is accidents. A good example is poly-tetrafluorethyene—PTFE. A DuPont chemist named R. J. Plunkett was working with tetrafluorethylene in cylinders. He wasn't aware that tetrafluorethylene could polymerize autocatalytically, and sometimes could detonate. One day in 1936 his assistant found that one of the cylinders that should have had 2,000 psi to 3,000 psi of tetrafluorethylene pressure in it, had no pressure at all. When the cylinder was opened, they found a white powder at the bottom of it. They examined the white powder. It was poly-C_2F_4. It lay fallow until 1943 when there was a critical need for a fluorinated material at Oak Ridge in the UF_6 diffusion plant. Teflon™ and its sister chloro-trifluorethylene were used for seals, gaskets, rings. It resisted the UF_6, which destroyed ordinary rubber.

Type X strategy is to establish a long-range central laboratory. An example would be DuPont's Eastern Laboratory in 1902. Another would be the establishment of a central laboratory in Mannheim by Bayer in 1890. It did nothing but work on new organic dyes. This was their response to the growth of the synthetic dye industry that began with Perkin's discovery of mauve dye in 1856. What DuPont did first in their central laboratory was work on black powder. For example, they substituted sodium nitrate for potassium nitrate. That was a revolution in the black powder business. It eventually led to the synthesis of ammonia. Before 1902 black powder contained potassium nitrate, charcoal, and sulphur. DuPont, in their new central laboratory, demonstrated they could use sodium nitrate. Once they had a big demand for sodium nitrate they went to Chile and got the waste from the guava from the gulls. They used that sodium nitrate to replace the expensive potassium nitrate.

Type XI is to exploit an invention that has much broader promise. An example has already been given in the discussion of fluorine. There one invention (the hydrofluorination of acetylene) led to a program to make refrigerants and propellants, and became a big business.

Another good example would be the synthesis of camphor at DuPont based on their original invention of special kinds of nitrocellulose. These are feed materials for synthetic camphor.

Type XII is acquisitions. That's a big subject. Acquisitions become a place where inventors get a good chance to work. Fresh people look at the business, and the buyers are frequently financially strong. Their staffs are usually more ingenious than that of the acquired company, and they are in a better position to attack the cost structure.

Some DuPont acquisitions were: (1) fancy dynamites; (2) Newport Chemicals was dyes. That was one of the best for DuPont; (3) They also acquired Grasselli, which made inorganic chemicals and organic acids; (4) They acquired cellulose acetate from the French; (5) They acquired photographic films from Pathe Film; (6) Synthetic ammonia was also acquired from the French; (7) Cellophane as well came from the French; and (8) they acquired rayon from the French.

Acquisition was a great strategy for them. Another one was: (9)

General Motors; which was different from the others, but it became totally dominated by DuPont. Then they stopped buying companies because of antitrust problems. Later they started again. I can find many more illustrations of the acquisition route and in lots of companies beside DuPont. The acquisition route is a major way to provide inventive people with opportunities for work.

Type XIII—cross-licensing is another excellent initiating strategy. This mostly concerns ICI and DuPont. I told you about the dye stuffs, neoprene cross-licensed to ICI, nylon cross-licensed to ICI, polyethylene came from ICI to DuPont, acrylic resins (which were a great, large-volume DuPont product) came from ICI to DuPont, ammonia came to DuPont, methylmethacrylate came to DuPont. A major kind of dye chemistry came to DuPont, as did white pigment: rutile TiO_2. The early TiO_2 was in the anatase crystal form, but the rutile form has 20–25 percent more hiding power.

GILMAN: A point that should be mentioned is you can't do cross-licensing if you didn't have a strong research staff, so that's one of the indirect, or secondary, values of a research operation.

KAZAN: Another strategy, *Type XIV*, comes from a capital commitment. If you make ammonia, you need hydrogen made by reacting water with hot coal (this yields a mixture of hydrogen and carbon monoxide). It didn't take very long for people to realize that the carbon monoxide could be used to make methanol and urea. Starting with the synthesis gas ($CO + H_2$) and controlling the composition ratios, and using an effective catalyst, methanol can be made. With other ratios and different catalysts plus some of the ammonia, urea can be made. So the commitment of capital to make synthetic ammonia brought urea and methanol, into the world. Developing and optimizing these processes provided 15 to 20 years of an inventive environment.

The market comes first in the *Type XV* initiating strategy. Acrylic resin is an illustration. There are not very many of these, but there are some. A variant is when there is a preexisting small market that opens up into something much larger. Examples: silk stockings, clear moisture-proof cellophane for packaging, and better tire cords. These are all places where a market was already defined, and it was decided to attack the market.

DuPont got into the fiber business via rayon and then cellulose acetate. That's why when Carothers started pulling nylon fibers out of his test tubes (or his workers did) DuPont was prepared to commercialize the new fibers. It already had spinning and drawing technology. It was already a rayon spinner.

Rayon made low quality hosiery, and the Japanese played marketing games with silk. When they wanted to, they made silk cheap, and *vice versa*. So there was a waiting market for the improved stockings that could be made with nylon.

In general, it's very difficult to take physicists, or biologists, or chemists and simply turn them loose. It is much better for some inventive people to be put into a program that says we're working on polyamides and polyesters; or something else. I think there are many inventive people from the physical sciences who function better with some program structure.

GILMAN: Yes, I'm sure that's true. On the other hand, with that approach really seminal inventions are unlikely to come. You won't get metallic glasses that way. Or the phonograph. A Tesla won't flourish in an environment like that, and Medawar[3] has made the point that minimal programming led to many of the seminal things in medicine.

Let's summarize to improve our perspective of these many strategies:

I —Grand scheme
II —Major theme
III —Star researcher
IV —Licensing
V —Fiat

VI —Purchase
VII —War needs
VIII —Bandwagon
IX —Accident
X —Establish central laboratory

XI —Broaden existing scope
XII —Business acquisition
XIII —Cross-licensing
XIV —Capital commitment
XV —Follow market

These are not completely independent of one another, but it is clear from this list that industrial research strategy is not a monolith. The variety in the list also indicates why research organizations are not easily understood, and why managing them is far more an art than a science.

Since there is no apparent pattern within this list of strategies, it is necessary to look behind it for general themes or recurring factors. Some are opportunism, flexibility, freshness, talent, entrepreneurial spirit, and unpredictability. The first of these can only operate if both an opportunity and an opportunist are present. Once they have met, that particular opportunity is gone forever. This emphasizes the future orientation of invention, and the folly of resifting old sands.

The story of Porter saying, "What's wrong with waxes?" illustrates the need for flexibility; opportunity can sometimes be created. It is a wonderful phrase because it implies so much and is such a marvelous reflection of the spirit of the inventor and the entrepreneur. If this doesn't work, try that. Hear! Hear!

The talents of individual people is a theme that is woven throughout the tapestry of the growth of the chemical industry. Actions, discoveries, and ideas of individuals dominated the critical branching points. Not committees, consultants, or constituencies. This point has been emphasized in the text by citing specific names of individuals instead of using pronouns, or collectives.

One of the purposes of using the anecdotal, or biographical approach, was to bring out the "doing" aspect because it is so important. To be successful inventors and researchers must *do*. Discussing, planning, presenting, thinking, and the like, may be helpful, but they are insufficient.

NOTES

1. D. A. Hounshell and J. K. Smith, *Science and Corporate Strategy—DuPont R&D, 1902–1980* (New York: Cambridge University Press, 1988).
2. *Ibid.*
3. See Chapter 7, herein.

7

LEGACY OF EDISON

It has been just so in all my inventions. The first step is an intuition—and comes with a burst, then difficulties arise. This thing gives out and then that—"bugs"—as such little faults and difficulties are called—show themselves, and months of anxious watching, study, and labor are requisite before commercial success—or failure—is certainly reached....I have the right principle and am on the right track, but time, hard work and some good luck are necessary too....

—Thomas A. Edison Letter to T. Puskas,
1878 The Edison Library Archives

How many of us could write numerous manuscripts, send them off to editors who would examine them very critically to determine whether they accurately described something original, and have 1,093 of them accepted? This is a record that has never been surpassed because no one else has had the unique combination of enormous energy, enthusiasm, and intelligence that Thomas Edison had. The manuscripts cited were, of course, patent applications.

Although the picture is often projected that he was a solo inventor, in fact Edison was the first great organizer of research work for industrial purposes. The outputs of his laboratories and companies were prodigious, largely because of his organizational skills. They covered several technologies of his day, including mechanics, electricity, materials engineering, and chemistry. The outputs reached beyond devices to novel, large-scale systems.[1]

Physical phenomena interested Edison, but mainly as a means to his ends: inventions. He discovered some such as the "Edison effect" (thermionic emission), and the piezoresistance of carbon black. The former became the basis of the first electronic device; the vacuum diode. The latter was used for many decades in the transducers of telephone sets, making Bell's rudimentary devices suitable for large communication networks.

Edison had a blindness that led to the commonly used malignment, "the Edisonian approach." Authors and writers say they will not use this very successful approach to complex problems with the implication that it is not sufficiently high-minded for their royal road. I notice that often they then go on to describe their own, less skillful, empirical approach. This critical phrase seems to have derived from Edison's blustering, uncomplimentary comments about theory and its practitioners. Consistent with his complex character, however, he employed a mathematician and others with theoretical backgrounds in his laboratories.

The extraordinary success of Edison's approach to the management of a research laboratory suggests that emulation of it is a policy that should not be dismissed. Critics have said that he was just a tinkerer with a flair for personal promotion. But anyone who has visited the Edison Memorial Site in West Orange, New Jersey knows that this is nonsense. The place is a monument to serious well-organized work. The man's office, as well as his napping couch, were in the large library where he was surrounded by thousands of technical books, which he studied routinely. The house where his wife and family lived were a short walk away from the laboratory site. He delegated well to talented strong associates, some of whom left his employment to become leaders in other organizations.[2]

Consider the industries that he had a major impact on: telegraphy, recording, lighting, telephony, phonographs, motion pictures, electric power generation, electric railways, electric storage batteries, housing construction, and others. He failed in his attempt to revolutionize mining, but the attempt was monumental. Although he dramatically improved goldenrod as a source of latex, his foray into botany was aborted by political economics. His many successes balanced against a few failures present a unique record!

It seems likely that his laboratories (Menlo Park: 1876–81, West Orange: 1886–1926, and Fort Myers: 1927–29) provided a model for the other industrial laboratories that were established in the latter part of the nineteenth century. They differed from his in being located within already existing corporate structures of the electrical, chemical, and communications industries, but their goals and operating modes were similar early in their histories. As time passed, they changed, most importantly through loss of their independence. It should be remembered that Edison created his own markets. No one commissioned him to invent the phonograph, or the high resistance electric lamp, or an effective motion-picture system. He commissioned himself.

It is easy to imagine a project review committee making short work of Edison's ambitious proposals. Vladimir Haensel wrote a fine facetious parable of this that I recommend to the reader.[3] It describes the fate of a proposal by him to work on incandescent lighting. His employer is a manufacturer of kerosene lamps.

Edison's best known invention is the artificial lighting *system* using incandescent lamp bulbs. After studying the electric lighting systems of the time, he concluded that a successful one would require high-resistance devices so they could be operated in parallel. But no one had been able to make a strong, high-resistance element that could be heated hot enough for adequate luminous efficiency. At the time this seemed to be a virtually impossible task. Starting with the knowledge of the high-temperature chemists of his day he put his staff to work trying various schemes: both compositional and fabricational. They tried a great many with eventual success. Does this sound familiar? It is intended to. There is a close parallelism between what Edison did and the "non-Edisonian" approaches of those searching for higher temperature superconductors in recent years.

The lamp bulbs were only part of what was needed. Adequate electric generators were not available, so they had to be designed, built, and tested. Inexpensive transmission lines had to be invented. Methods for connecting the bulbs to the system, as well as putting wires safely into houses, and many other details had to be worked out.

Even after Edison had built and demonstrated working equipment for a particular technology, no army of investors or users with cash in hand appeared. Instead, the struggles began. To find financing, to fight the claims of competitors, to persuade potential users of its merits, to establish manufacturing operations, and so on. The advantage to Edison of enduring these struggles was that he was able to finance himself; albeit with vast swings in his income from millions per year to large deficits. Self-financing preserved the independence of his organization, and Edison understood the importance of it quite consciously. Its importance derives from the nature of research activities, and the people who execute them.

In these times of concern for the productivity of research organizations it seems to me that contemplation of the ways in which Edison operated is worthwhile, especially when his ways are contrasted with the ways of some of our contemporary organizations. From his laboratories sprang whole industries, some of which Edison owned and operated, although business was a means to an end for him—not a primary interest. The profits tended to disappear into his next enthusiasm, and his attention to business tended to wander, creating opportunities for others to take control.

Research organizations are intrinsically different from organizations engaged in production, marketing, or finance. Many of their day-to-day activities are the same as those of other functions, but these are not the central ones. To paraphrase Edison, it is inspiration that is essential, not perspiration, even though inspiration occurs rarely.

What did Edison organize for? He organized, I think, for making efficient (that is, educated) sequential searches for the solutions of what he perceived to be important technological problems. He searched through very large "trees" making selections at the branch nodes. Sometimes he got onto branches unintentionally and made unexpected discoveries, which he lingered over. If the territory was particularly unknown to him, he also did some random-walk searching along the lines of the theory presented by Montroll.[4] Random-walk searching is very inefficient, so given the smallest of clues Edison began sequential searching.

How did he organize? He gathered groups of specialists. Many of them were artisans such as machinists, glassblowers, mechanics, and modelmakers. He needed to leverage his own abilities, but he also employed people with abstract knowledge who had been schooled in science and engineering. At Menlo Park, the scene of his greatest triumphs, they worked in large loftlike rooms so the operating mode must have been highly interactive. There were foremen, but little other formal organizational structure. Thus the administrative staff was held to a minimum, 5 percent or less.

As for long-range planning, Edison spoke of his "intuition," not of his "plans," when he was speaking in the context of making inventions. He did do planning, of course, when he was working in the context of development or commercialization work. Once an invention has been made the context shifts dramatically from the unknown to the known. If a decision is then made to develop a prototype, detailed planning becomes highly appropriate. Judging from the literature on research policy and management, this qualitative difference is rarely accepted. The result is low productivity in the arena of invention, and other intrinsically creative activities.

With an output that cannot be described initially because it does not exist, how can rational purposes, strategies, and tactics for an organization be described? Poorly, is the only honest answer. With Edison's rule regarding the ratio of perspiration to inspiration (99:1) in operation, how can anyone be convinced that a research group is adequately productive? Past performance, not promises of the future should be the basis. The difficulties associated with explaining why an organization that fails most of the time (by conventional standards) should be financially supported anyway is one of the reasons why Edison struggled for independence for himself, and for others.

When he had the opportunity at the end of World War I, Edison argued for an independent operating mode for government research; in particular for the Naval Research Laboratory. He failed.

During World War I, Edison served on the Navy Consulting Board, an impressive group of inventors and scientists that provided technical advice to the U.S. War Department. From this platform he promoted the idea of establishing a Naval Research Laboratory. He fought to have it placed under civilian control and be located away from Washington, D.C. He wanted it near New York in order to give it independence from the Navy's *status quo*. He did get it established, but squarely under the thumbs of the admirals on the bank of the Potomac at Washington. Although it has lacked the independence that Edison envisaged, it has served the Navy well. Think of how well it might have served if the master's advice had been followed!

Even the walls of the traditional citadels of independence have been breached in modern times. Namely, those of the universities. This has come indirectly through fiscal policies that have slowly, yet surely, extracted independence from their research activities. Patrons have always had a strong influence on university operations, but not so blatantly as in recent decades. The change and its probable consequences have been described eloquently for the British scene by Sir Peter Medawar:[5]

> the government was engaged in reorganizing the funding of the Research Councils on the basis of the retail trade: customer-contractor was to be the principle, in which the scientist was to put up a research proposal to the government and if the government approved, the scientist would be given a contract to undertake it and bring about the desired result. This was a very bold innovation and had not been the basis on which research had hitherto been conducted. Nor was it the proceeding that had given us penicillin, insulin, the discovery of the blood groups, the elucidation of the causes of myasthenia gravis, the transplantation of tissues, or the discovery of the genetic code. Scientific discovery cannot be premeditated.

Medawar describes what happened in Great Britain a decade or two after it had already happened in the United States. The "bold innovation" that he describes had the effect of weakening the responsibility ethic in the scientific community by reducing the independence of the individual scientists. This was a predictable effect. When the taste of the artist is replaced by the taste of the client, disappointment follows.

The transfer of responsibility from individuals to various collectives is the most central management trend of the past few decades. It is debilitating because it replaces specificity with diffuseness. It is the diametric opposite of the approach of the most productive of research organizations: Edison's. Responsibility for the output of scientific papers, inventions, and development prototypes has been gradually transferred from individual workers to "reviewers." That is, to peers, to group leaders, to department heads, to deans, to vice-presidents, to advisory committees, to contract monitors, to politicians, and to trends promoted by news media. Who can exert grass-roots management of people in such a system? No one can. Therefore, abstract financial management is gradually becoming predominant.

When people are given individual responsibility, most take pride in it, and exercise it. When it is transferred from them, pride in it is lost, it becomes diffuse, and degenerates. People who are held individually responsible, and thereby allowed to feel like entrepreneurs within their own domains, tend to be most productive. They compete with one another for attention and rewards.

Starting about 20 years ago, most research laboratories have been balkanized, largely through fiscal policies. That is, they have become divided into collections of projects rather than groups of people. Through much talk, and in some cases the reality, of zero-base budgeting, insecurity has become endemic. Insecure people who identify with projects rather than their own egos cannot be expected to be highly productive. Nor can they provide desirable role models to youth.

In the beginning of the balkanization process, most people did not realize it was happening. As the fiscal noose tightened, the process became apparent, but no means for stopping or reversing it was available. Then pandering began. Pianissimo at first, followed by a crescendo, until recent years have exhibited fortissimo events. Among these have been a variety of pork-barrel projects of dubious merit, substitute energy projects of negative merit, misplaced biotechnology, the ceramic automobile engine follies, fantastic space warfare, and bloated superconductivity enterprises. If common sense prevailed these would have been hooted off the stage. Each has had a kernel of merit, but has been promoted to an untenable position by pandering to greed and naivete.

At the same time that these aberrations were occurring, so were many marvelous discoveries and inventions being made. Thus the issue is not one of absolutes. Rather it is one of efficiency, or effectiveness. That is, high ratios of outputs to inputs. An obvious improvement would be to not waste resources through hyperactivity devoted to temporary fashions. Also, not waste them by large oscillations in support levels. Fashions in science are unbecoming. Invention is where you find it, not where it is decreed.

The art of making critical choices was a great strength of Edison that seems to have become diluted in recent times. Probably because the goal is often not to solve a problem, or find an interesting topic for investigation, but to ingratiate a source of finance. This has lead to much activity but has diluted and diffused significant results. The nature of the problem was described superbly by another great man of the nineteenth century:

> *Invention consists in avoiding the constructing of useless combinations and in constructing the useful combinations which are in the infinite minority. To invent is to discern, to chose.*
>
> —Jules Henri Poincare (1854-1912)

The lessons of the Edison legacy are several. One that is abundantly clear is that organized technical research pays. Provided it is managed wisely, using modes that are appropriate to it—not ones appropriate to other activities. Another is that research workers who exercise responsible independence have been, and remain, crucial to high inventive productivity. Also, high levels of activity are important, but must be directed by critical judgments. Finally, a substitute for genius remains to be found.

NOTES

1. M. Josephson, *Edison* (New York: McGraw-Hill Book Company, 1959).
2. W. B. Carlson, "Thomas Edison as a Manager of R&D: The Case of the Alkaline Storage Battery," *IEEE Technology and Society Magazine* (December 1988) p.4.
3. V. Haensel, "Lucky Alva," *Research Management*, Vol. 10 (1967) p. 135.
4. E. W. Montroll, "Dynamics of Technological Evolution—Random-Walk Model for the Research Enterprise," *Proc. Nat. Acad. Sci.*, Vol. 76 (1979) pp. 6030–34.
5. P. Medawar, *Memoir of a Thinking Radish* (Oxford: Oxford University Press, 1986).

8

FOLLIES AND THEIR REDUCTION

There is no reason for any individual to have a computer in their home.

—Kenneth Olson, President,
Digital Equipment Corporation at
World Future Society meeting, 1977.
(one million were installed by 1982).

When I emerged from graduate school in the early 1950s, I was offered a job at what was called the Aircraft Nuclear Propulsion project. It seemed exciting and full of promise to me at the time. Little did I know that it was one of the largest technological follies in history—perhaps *the* largest in the United States up to that time. Fortunately for me, I had another offer and went to work at the General Electric Research Laboratory instead. Others wasted very large amounts of their time and taxpayers' money on a fruitless endeavor (14,000 people in seven states for over 15 years).[1] This was my first, but in no way last, encounter with technological follies.

By a folly is meant a project whose failure could be prophesied with considerable certainty from the start. Thus follies are to be distinguished from projects that no one would consider to lack feasibility initially, but for some reason turn out not to be feasible. The distinction is sometimes subtle, but careful study can usually put a given case on one or the other side of the border. For example, following the discovery of the Carnot cycle for heat engines, there has been no sensible reason for pursuing a project whose goal is the construction of a perpetual motion machine. In other words defying established technical principles is folly. Similarly, defying established economic principles is folly.

In the case of the nuclear powered aircraft there were at least two technical reasons for not starting a large project directed toward building one. These should have been (and probably were) apparent to good experienced engineers of that time. One is that the power flux (watts/m^2) that could be extracted safely from a lightweight nuclear reactor would not be adequate for aircraft propulsion. The required power flux is at the upper limit of what internal combustion engines can supply, which is much higher than that for nuclear reactors. The other is that even if the first problem could be solved, the weight of the shielding needed to protect personnel from the reactor's radiation would make the thrust-to-weight ratio much too low.

Notice that these reasons are quantitative. This is important because it is almost always the quantitative aspects of a proposed innovation that determine its fate. But those who delight in quantification are a small fraction of the population indeed. Few people want to interpret the statement that "the watts per square meter per kilogram are too low." It requires real mental effort. Whoever raises the issue becomes labelled a spoilsport, while someone who says with some dignity, "don't worry, some research will solve the problem" becomes a hero. In the meantime, staggering amounts of money are wasted. Not only is the money (capital) spent without a productive result, but also it is not spent in some way that might be productive. Thus it is lost twice over.

The subject of follies is pertinent here because they are a form of negative inventivity and thus reduce the overall average. Reducing them can raise the average just as conservation can reduce average energy consumption.

An occasional folly might not be very disturbing. But they abound. At about the same time that the Aircraft Nuclear Propulsion project was active there were others of similar ilk: Rover (nuclear powered rocket), Pluto (nuclear powered ram-jet), and Orion (spaceship powered by small nuclear explosions). This last was mainly a paper study.

Dozens of commercial follies are listed in a recent book,[2] and in an article.[3] Only a very few can be discussed here. Such follies might be expected to be a "normal" part of the transactions of a society. They would arise, spend some time at center stage, and then fade away. However, the largest of them seem to be consuming ever larger fractions of the commonweal. Therefore, it seems worthwhile to try to identify some of their common characteristics in the hope that this will help in avoiding the worst of them in the future.

Follies are prophecies. Flawed prophecies. It may be that the prophecy itself is flawed, or that the prophet is flawed—or both. The flaws take various forms: violations of physical principles; neglect of economic principles, or poor timing. In some cases, follies are absolute nonsense. In others, they are prophecies that have been puffed up to unrealistic proportions. Some violate technical principles. Many violate economic principles. In addition to those previously mentioned, some examples are: X-ray laser weapons, molybdenum turbine blades, thermoelectric refrigerators, Corfam, superconducting batteries, photoelectrolytic hydrogen, oil from algae, life in a test-tube, superconductive levitation of railways, photovoltaic central stations, controlled-fusion power stations, materials processing in space, ceramic aircraft turbines, and many more.

I have tried to avoid cases of simple failure. That is, cases where the outcome could not have been predicted with the information that was available at the beginning. I think those I have selected could have been avoided through the use of a little hard-headed forethought.

Consider as an example the synfuels project during the "energy crisis." From 1974 to about 1982, the government and various industrial firms spent large amounts of money annually (billions of dollars) on processes such as coal liquefaction, coal gasification, and the extraction of oil from tar sands. The premise was that although these are expensive processes, the rising price of crude oil would soon reach levels that would make them economic. Prices were projected to reach 100 dollars per barrel by about 1985. Even if such prices had been reached, synfuels would not have become economic because the costs of building and operating synfuels plants were escalating in synchronism with the rising crude prices.

To show this synchronism, both crude oil prices and estimates of the costs of synfuel plants can be plotted on the same piece of paper (Fig. 8.1). It is clear that the two trend lines are essentially parallel, not convergent. They do converge slightly, but their intersection point lies far into the future—at least 55 years. Thus there was no reason to expect that synfuels were becoming competitive.

What was wrong during that period? Apparently, no one did time-dependent input-output analyses of the effects of high oil prices on the whole economy. In other words, they did not project the secondary responses that must occur following a major perturbation of the economy. This point must have been obvious to any well-educated economist. It seems curious to me that they did not speak up. Perhaps they did, but the government didn't listen. Also, the more astute among the chemical engineers of that time must have understood the price and cost dynamics of the situation. Where were their voices? Or were they unheard? The multibillions of dollars that were wasted could have served better purposes.

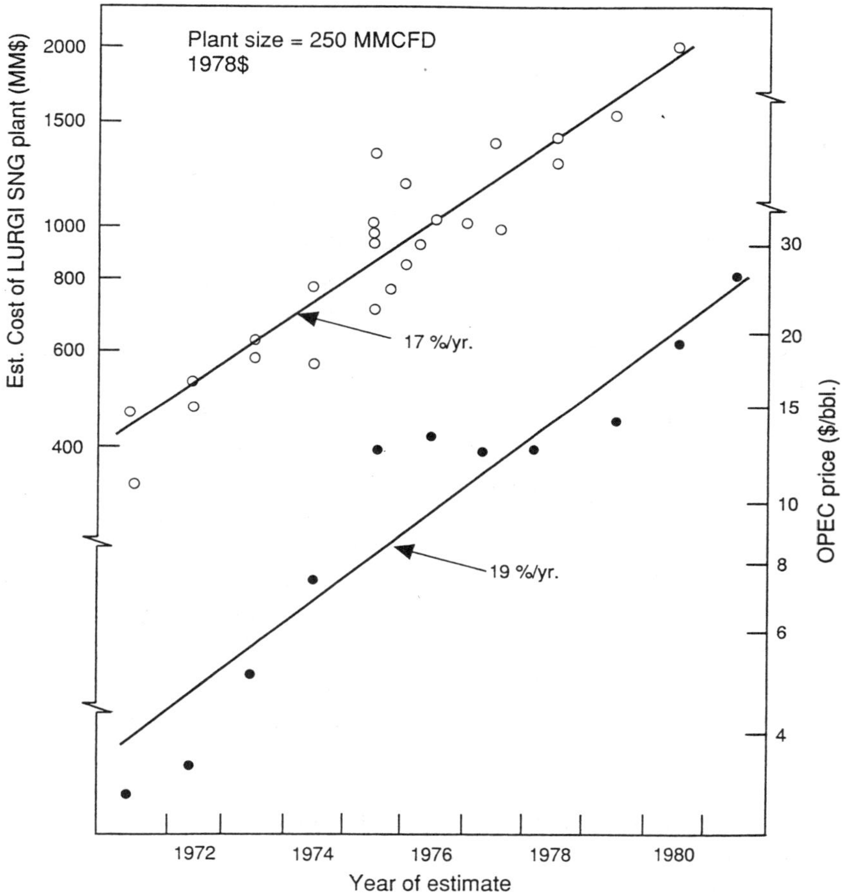

FIGURE 8.1. Costs of constructing synthetic natural gas plants compared with posted crude oil prices for the period from 1971 to 1981. The plant design is constant, being a particular size (250 million cubic feet per day) from a particular designer (Lurgi).

Toward the end of the 1970s, and through most of the '80s, another pathology arose. This was the craze for biotechnology. It grew from some of the greatest scientific discoveries of this, or any other century. Nevertheless, it became a folly because its promoters ignored known technical and economic constraints. Assertions were being made that virtually all chemical processing could be done more effectively using biological organisms than by conventional methods. Soon to be revolutionized were agriculture, mining, energy production, food processing, and medical practice. This caused a stampede of unwary investors.

My interest in it was that of a casual observer, until it became necessary for me to formulate an opinion when I was given management responsibility for a group of biotechnology researchers. Before that I had felt vaguely uneasy about what I was reading in the newspapers about the field, but I hadn't formed an independent opinion.

However, I had come to respect demand charts as planning tools, so I made one for biotechnology as described in Chapter 3. Following the reasoning given in that chapter, I concluded that most of what was being said in the business literature didn't make good sense. I was led to the conclusion that realistic research and business strategies in biotechnology had to be considerably more sophisticated than most people were contemplating at the time.

Although I think that industrial biosynthesis has a brilliant future, it does not lie in competing with conventional industrial organic-synthesis.

The best people fell into the biotechnology trap. Not just the small understaffed companies run by incompetent people with bad motivations, but also companies that are well financed, with plenty of competent people who spent much time developing judging rituals. Companies with good management, good marketing activities, and creative people reinforcing every department.

The science underlying the biotechnology craze was revolutionary. It has changed the world in major ways, and will continue to do so. So what went wrong? Probably a lack of quantitative thinking on the part of the majority of the people involved. Cost and time scales seem to have been unrealistic. A lack of appreciation of the time value of money on the science side, and a lack of understanding of the intricacies of engineering practices on the business side.[4]

The DuPont company employs many very bright people, both engineers and scientists. Yet they were razzle-dazzled by biotechnology. In one specific case, they announced that they would find ways to fix nitrogen (remove it from air and convert it to various chemical compounds) in large quantities through genetic engineering. A fine goal indeed. At one time they had a four-pronged overall program that they committed about a billion and a half dollars to. Because of the intricacies of biotechnology, which stretched time scales and therefore costs, it has come nearly to nought.

One of the keys to how companies become embroiled in follies is that those who know better do not speak up out of fear of losing money or power. Another interpretation of the phenomenon is that expertise has become too compartmentalized. So, for example, an expert on biological nitrogen fixation knows too little about engineering constraints, project schedules, or finance.

But this doesn't excuse the boss. In a society that is money oriented, people's incomes are determined by how wise they appear to be, so the truly knowledgeable are silent. They don't disturb the fools.

How does an industrial organization decide in a broad way what topics to work on—and not get snookered? One purpose of a technical staff is to take inputs from throughout the world, analyze them, and recommend which ones are most important for the company's strategy. But it is rare that the procedure is followed to completion. In particular, it is rare that there is adequate consideration of what might happen if a particular input led to a technical success. What would the new product or process actually be? Not its general characteristics, but its commercial specifications. What would be off-loaded from the shipping dock? In other words the overall *system* does not get analyzed. Researchers usually consider such analysis to be a threat to their interests. I think that it in reality it strengthens their case, assuming they have a case. If they don't, they won't get long-term support anyway. Early knowledge of problems should help in solving them. It is hard to believe that ignoring or hiding problems will smooth the path to eventual success.

In the early stages of the development of a technology it is obvious that risk is very high. But, for the person who judges everything correctly, the returns are very large. For example, a company that judged things correctly in the biotechnology arena, was Applied Biosystems. They saw an opportunity for new tools. This was not a very sophisticated judgment because the early stages of any type of technical development starts with the tools. But it was wise. In an earlier era, a similar example is that of Hewlett-Packard. It started by making tools during the early days of industrial electronics. Note that Applied Biosystems was started by former Hewlett-Packard employees.

The discovery of ceramic "high-temperature" superconductors started another craze. There was talk of using them to levitate railroad trains; to revolutionize the generation of electric power; to transmit and store electric power; to propel ships; to provide power for automobiles; and to accomplish other great feats.

As an example of the folly of these dreams, consider superconducting batteries (power storage). It is another example of how quantitative reasoning helps to avoid pitfalls. A high-performance battery with zero internal resistance sounds great. A nitpicker might ask about the alternating-current reactance, but let that pass. Instead, lets ask what such a battery would do that a chemical one won't? The short answer: not much. Why? Because the superconducted currents are macroscopic. They obey Faraday's laws, exerting macroscopic forces on themselves and the materials that contain them. This means that the energy densities in the magnetic fields that surround the currents must be balanced by strain-energy densities of equal magnitudes in the materials that confine the currents. Using optimistic strengths for commercial ceramics, the first problem is that such a battery would be monstrous in size in order to obtain useful capacity.

Suppose an inexpensive method could be found for making superstrong ceramics. The best that could be made would support a strain-energy density of the order of the energy densities of chemical bonds. But that is how ordinary electrochemical batteries store energy, in chemical bonds. So, after spending much money, time, and effort on the development of a superconducting battery, at best one will end up with the equivalent of a high-performance chemical battery.

But lets not give up yet. Suppose there were a pressing need for a battery with very low internal resistance, so we decided to go ahead with our development project anyway. What would the battery cost?

Are there any fundamental limitations on the costs of ceramic processing? The answer is yes. There are either one or two, depending on the composition. Silicates are often cheap, but the rare earth oxides needed for superconductivity are not. There are two reasons for this. First, although the average abundance of the rare earths is about the same as that of copper, there are no appreciable concentrations of the ores. Second, the rare earths are expensive to separate and purify because they are all very similar chemically, so chemical reactions do not discriminate among them, and physical methods are slow.

The other limitation is that, unlike metals and polymers, ceramics cannot be shaped by means of plastic deformation. Their shaping requires mass diffusion, which is inevitably comparatively slow. Mass producing ceramic parts does not lead to low costs because of the relatively high rent that must be paid for the capital equipment involved. For example, the aluminum oxide wafers used in large quantities as semiconductor substrates cost 30 dollars per pound, or more. A demand chart, such as Figure 8.2, for structural materials indicates that this price level corresponds to the uncommon metals, Nb, Ta, Be, or cemented carbides. All of these are too expensive to be considered for the construction of large civil engineering structures.

The same problems exist for other large-scale applications of oxide superconductors such as high-field magnets, levitated trains, and so on. These problems, however, do not limit applications in which small quantities of the superconducting material is an integral part of a device. In the area of microelectronics there may be many applications for them.

What do these various technological follies, and others like them, have in common? Some of their common characteristics are as follows:

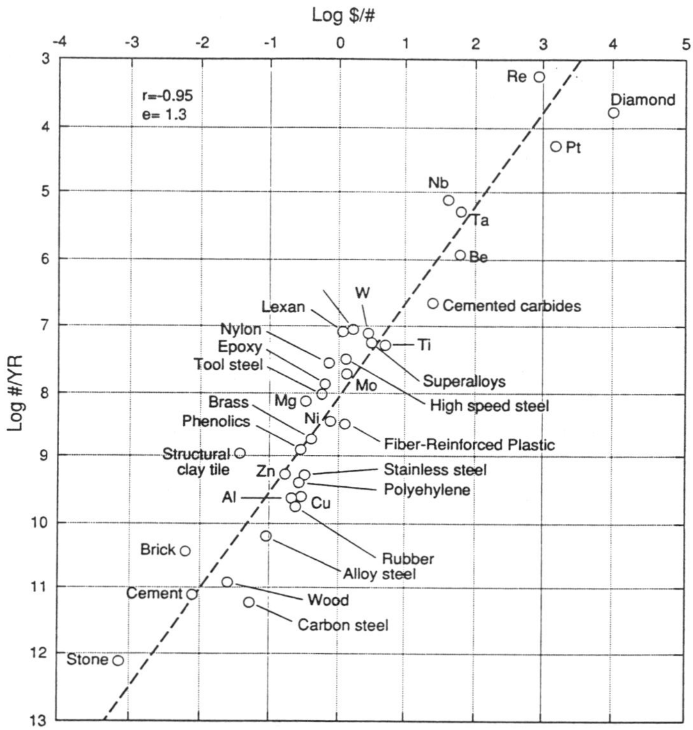

FIGURE 8.2. Demand chart (price versus consumption) for structural materials (courtesy J. H. Westbrook). Note that the chart is rotated 90 degrees from those of Chapter 3, and r = correlation coefficient, while e = elasticity parameter.

1. Success of the technology would fulfill a dream. This is not a distinguishing characteristic, however, since many valid technologies fulfill dreams.
2. The virtues of the technology are treated with awe and sometimes magnified beyond common sense. The faults tend to be ignored.
3. There is a call for new research support, or it is proposed to proceed directly to demonstration projects. The latter are usually fragmented, so audits of the overall technology system are difficult, often impossible, to do.
4. The flaws in the technology can only be recognized by those with a complete knowledge of the pertinent parameters. They tend to remain silent.
5. One or more critical parameters are omitted in discussions by its protagonists (promoters). Most often this is cost, but it may be a

technical parameter such as poor stability in humid air, or poor resistance to radiation, or brittleness if cold, etcetera.

6. Antagonists (critics) are dismissed as spoilsports (dream busters).
7. Criticisms are met with ad hoc excuses, or parried with promises that the flaws will be rectified by future developments.
8. Much attention is given to overcoming some faults, but little or none to the fatal ones.
9. The pro–con ratio starts high (often greater than one), and decays exponentially. In contrast, for healthy technologies, the pro–con ratio often starts near zero, and rises exponentially.

Some of these are similar to those involved in pathological science[5].

The overall picture that emerges is that the protagonists of new technologies often feel that they-must not reveal, or admit, that their prophecies have any flaws. This may be effective salesmanship in the short term, but it does not lead to success in the long term. There are so many wonderful opportunities for technological advances that it would be far better to admit the existence of flaws and either correct them or move on to new technologies.

NOTES

1. J. Tierney, "Take the A-plane," *Science 81* (Jan./Feb. 1981) p.46.
2. S. P. Schnaars, *Megamistakes—Forec the Myth of Rapid Technological Change*, (New York: The Free Press, 1989).
3. J. Hecht, "Boondoggles We Knew and Loved," *New Scientist* (December 23, 1989) p.32.
4. R. Teitelman, *Gene Dreams: Wall Street, Academia, and the Rise of Biotechnology* (New York: Basic Books, 1989).
5. I. Langmuir, "Pathological Science," Transcribed and edited by R. N. Hall, *Physics Today* (October 1989) p.36.

9

TIME THAT IS NEEDED AND ITS COST

No great thing is created suddenly, any more than a bunch of grapes or a fig. If you tell me that you desire a fig, I answer you that there must be time. Let it first blossom, then bear fruit, then ripen.

—Epictetus, Discourses,
Book I, Chapter 2.

A virtually universal human failing is underestimating the time required to execute a project. This seems to be true of even simple activities, but it is compounded for complex ones, such as writing a book, or doing some research. It applies to both personal tasks and to the collective behavior of groups. This is why backstage on the opening night of a play is "panic city." This is also very often a source of substantial extra costs during the introductory period of a new product or process. These extra costs may occur if production facilities are built too early, if too much inventory is collected, if a sales force is hired too soon, and so on.

DIFFUSION OF TECHNOLOGY

Although there is no really accurate way to coordinate an innovation schedule, estimates can be improved by studying, or at least being aware of, past experiences. From these experiences it is known that new technology spreads by "diffusion"; not by propagation. That is, it does not spread along direct paths like acoustic or optical waves. Rather, it spreads through a complex series of nearly random steps like the diffusion of a concentration gradient. Some of the steps advance its cause, others retard the cause. The ratio of advance to retardation in a given amount of time depends on the size of the economic driving force for the innovation. The vector sum of the many individual steps leads to net diffusional spreading.

Innovation diffusion occurs in both time and space. From the place where the technology is first discovered and developed it diffuses to other places. When it reaches a new place its use begins to grow diffusively at that place. The growth is known to follow the simplest of kinetic rules. Namely, the rate of growth is proportional to the amount of the new technology that is already being used at that place. The consequence of this simple rule is that the amount of use grows exponentially with time. Thus the amount grows very slowly at first, but later very rapidly. On the other hand, the fractional growth per year is constant.

The speed with which diffusion and subsequent growth occur depends on the net driving force, and this is determined by the economics of the situation. The primary factor that determines the speed is the economic utility created by the new technology. This must be relatively large because there are always countervailing factors. One of these is the cost of the investment needed to introduce the new technology. The larger this is, the more competitive projects there are seeking the same investment funds. Factors that are difficult to quantify are also important, such as the enthusiasm of the innovators, providing this enthusiasm is based on reality.

It is important to recognize that economic utility is not identical with technical utility. A new device or process may give improved performance relative to older ones and yet not increase economic utility. Initial cost may be the problem, or high maintenance costs, or other factors. Even if the new item has a small net increase in utility, the force driving its introduction to the marketplace may be too small to give it an adequate diffusion rate. A large-scale example is that of nuclear power plants replacing those fired by coal, gas, or oil. Although the nuclear plants are cheaper to run, the difference is only marginal, so the driving force for their introduction without subsidies has been slow and they have been vulnerable to political attack.

It is also important to recognize that economic utility is not necessarily the same as monetary value. For commodities, like steel and wheat, the two are closely related, both qualitatively and quantitatively, but this is not true in general. Quantifying the utility of safety, for example, is very difficult and can only be done in some statistical sense. Fashion is also difficult to quantify. As are luxury and health.

Another complication associated with quantifying utility is that it is context dependent. For example, consider the transition from black-and-white to color information displays. In the entertainment field (television) this transition occurred quite fast (eight years), indicating that a large difference in utility was driving the transition. In the business field (alpha-numeric and graphic data) it occurred more slowly at approximately the same difference in hardware cost. It appears that relative changes, as well as absolute changes, are important in determining the rates at which transitions occur in markets.

The value of money, in addition to the value of the entities it represents, declines with time in a well-defined way that is determined by real interest rates. This creates the imperative that, in order to be widely used, new technology must have some critical amount of increased economic utility. If it has less than this critical amount, it will not be adopted because the pace of introducing it will be slow, and therefore the real monetary cost will be too high. Another way of saying this is that the introduction time will become infinite.

MARKET PENETRATION CURVES

Even when the driving force is adequate, the length of time needed to introduce a new technology surprises most people. In order to define it, consider Figure 9.1 showing a schematic penetration curve. Such curves have a simple analytic form, and they are followed remarkably well in a large number of cases in which an existing technology is replaced by a newer one.[1]

Changes in virtually all complex systems occur in the same way from a kinetic viewpoint. A new entity begins to replace an older one. At first the rate is slow and proportional to the amount of the new entity. As the amount of the new increases, the rate increases until it reaches a maximum when about half of the domain of the old has been taken over by the new. Then impingements, or other phenomena begin to reduce the rate of growth of the new until the rate becomes zero, and the process becomes complete.

The curve of Figure 9.1 has a variety of names, which have resulted from its various applications. Some of these are the growth curve, the logistic curve, the "S" curve, the sigmoidal curve, Pearl's curve, Verhulst's curve, Gomperz's curve, and others. Readers interested in the mathematics can find discussions with increasing generality in Gilman,[2] Fisher and Pry[3]; and Montroll and Badger;[4] respectively. Some discussion is also given in Appendix IV. For an overall review, see Hurter and Rubenstein.[5]

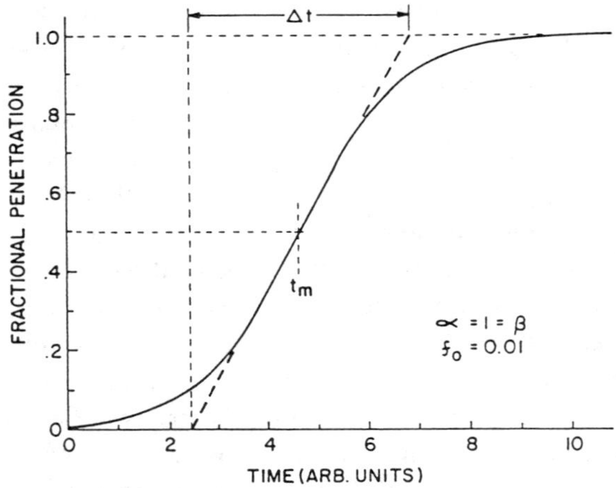

FIGURE 9.1. Schematic market penetration curves. The S-shaped curve is the actual form that is followed as the market fraction increases from zero to one. The dashed line is a linearized version. The overall penetration time is taken to be t. It may be seen that this is the time needed for the market fraction to change from about 10 percent to about 90 percent. The mid-point time, t_m, indicates the historical time period during which the penetration process has occurred.

Market Penetration Times

The dashed line in Figure 9.1 represents a linearized version of the sigmoidal curve in order to define two characteristic times. These are the overall penetration time, *T*, and the chronological time, t_m at which half of the penetration has occurred. The latter is mostly of historical interest, while the former gives an indication of the length of time that has been required for various substitutions to occur in times past. This provides a guide to the times that are likely to be needed in future cases.

Data for 60 cases is listed in Appendix III. These substitutions occurred during the past 100 years, approximately. Some took a long time, such as the use of oil rather than coal in the propulsion of boats (58 years). Others occurred within a few years. Transient cases such as novel clothing designs, or faddish toys, are not included. Penetration curves are not necessarily independent of one another; two or more of them may compete. For example, someone who buys a laptop computer is unlikely to also buy a desktop machine.

Figure 9.2 presents the penetration/time data of Appendix III as a histogram with five-year intervals. The median interval is 16 years, while the average is 22 years. These are long times compared with the dreams of entrepreneurs,

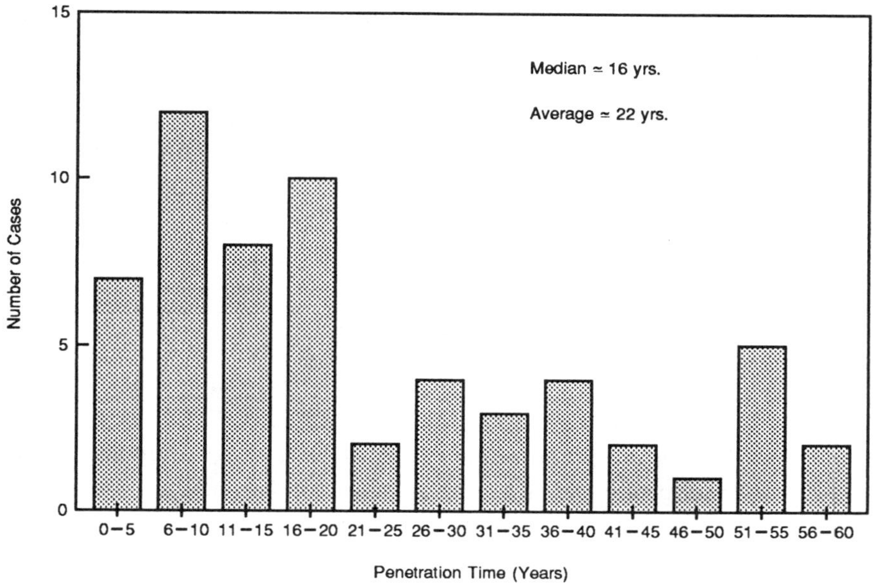

FIGURE 9.2. Histogram of penetration/time data listed in Appendix III. The median of the data lies at about 16 years. The average is 22.5 years. It is apparent that no simple distribution law will describe the data. However, the more complex the substitution is relative to the previous technology, the longer it takes for penetration.

but are realistic for planning purposes. Intuition suggests that shorter times are possible, probably because stories of fast successes tend to be remembered preferentially. However, the data indicate that intuition is unreliable for estimating penetration times.

The two major cases in which I have personally had involvement did adhere to the median of the histogram: about 16 years. The first of these is directionally solidified components for aircraft turbine engines. The inventor was Frank VerSnyder. My participation was as a consultant to him at Pratt & Whitney Aircraft. The technical advance is to cast components such as turbine blades and vanes in such a way that the boundaries between the grains of the castings are aligned with the principal stress axes of the components. This minimizes the stress across the relatively weak boundaries when an engine is operating. By improving the strengths of the components, a substantial improvement in engine performance is realized. The idea is simple, but it has led to a major improvement in engine performance.

Laboratory work on this idea began in about 1961, but it took until the mid-1970s for this type of engine component to appear in the engines of commercial passenger aircraft. During the interim, after the idea had been demonstrated to be effective, extensive testing had to be done, as well as the development of effective manufacturing methods.

The second case is that of magnetic metallic glasses. The inventor was Ho-sou Chen. My involvement was as the director of the Materials Research Center at the Allied Chemical Corporation where the invention was made. Metallic glasses are made by solidifying special alloys so fast that they do not have time to crystallize; thereby obtaining a "glassy" material. These metallic glass materials are magnetic if they contain magnetic metals such as iron and cobalt. Moreover, they are more easily magnetized, and demagnetized, than any other materials. This makes them very useful as the cores of transformers and other magnetic devices in which the direction of magnetization changes rapidly. They are now being used in large quantities to make premium power transformers. The invention was made in about 1970, and large-scale commercial production did not begin until about 16 years later. The time needed for these materials to substitute for 50 percent of the power transformer market will be even longer.

It is natural to wonder why it takes so long for market penetration to occur. The formal answer is that the initial rate of penetration is proportional to the amount that has already occurred. But this amount is initially very small, so the rate is very small. Another way of saying this is that customers must be "sold" one at a time until they begin to "sell" one another. In still other words, the selling process is one of diffusion rather than propagation.

Financial Characteristic Times

The commercial introduction of new developments involves several characteristic times. One of these has just been discussed. Another is the reciprocal interest rate, which represents the time associated with the value of money. If the interest rate is, say 8 percent per year, the reciprocal is 12.5 years which is the half-life of the value of the money, and is a measure of the time that one can afford to wait for a new development to come to fruition. The higher the interest rate, the shorter the time that can be waited.

Still another characteristic time is the reciprocal inflation rate. And the reciprocal growth rate. For planning purposes, the time during which the price of a product decays is also important. It is not obvious that this is a characteristic time so some words of explanation will be given here.

After a new product has been brought to the marketplace at a particular price, the price begins to change. Almost invariably the price decays to a lower value, eventually reaching a stable level (after correction for inflation or deflation). The decline in price results from various factors. One is that the manufacturing costs decline as experience and volume are gained. Another is that the responses of competitors tend to drive price reductions. These factors eventually lose their effects as the market becomes saturated. Then a stable equilibrium tends to be reached.

Price declines follow a simple behavioral law. The law is an exponential one that is statistically based. The declines tend to be relatively large at first, becoming smaller and smaller as the stable level is approached. Since the declines follow a specific analytic law, they can be described by means of a single characteristic time. The advantage of this is that once the initial behavior is known, the rest of the behavior can be predicted.

Let the initial price be P_o, and the eventual stable price be P_s. Then the total price decline will be the difference, or $P_o - P_s$. If the analytic form of the behavior is exponential, the rate of decline is proportional to the remaining difference between the current price and the stable price. Thus, if P is the price at any time, so $P - P_s$ is the remaining difference, the rate of decline will be:

$$dP/dt = -D(P - P_s)$$

from which the price-decay law can be obtained by integration. It is as follows:

$$(P - P_s) = (P_o - P_s)\exp(-Dt)$$

Taking logarithms of both sides of this equation yields the following linear equation:

$$\log(P - P_s) = \log(P_o - P_s) - Dt$$

which is a straight line when plotted on semilog graph paper. This provides a means for testing the price-decay law, and determining the characteristic time, $1/D$, for the price decay. D is the slope of the line.

A classic case is that of the Model-T Ford automobile because the product remained essentially unchanged for a long period of time. Figure 9.3 shows the data plotted semilogarithmically. The data extend from 1909 when the Model-T was introduced until 14 years later when it was discontinued in 1923. The introductory price was \$3,300, which declined to \$980 (in constant dollars) at the end of the period. The best fit of a straight line to the data is

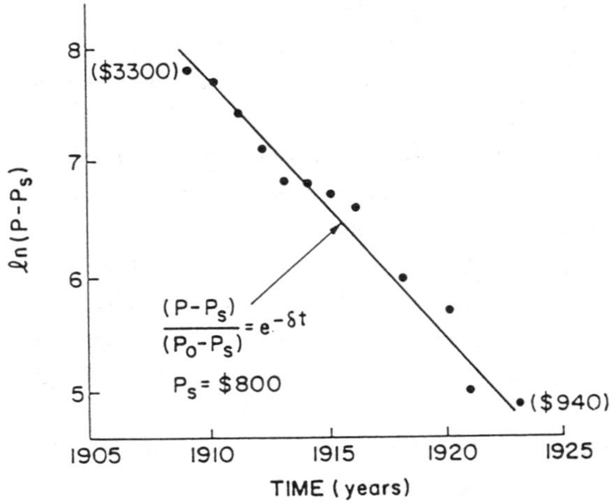

FIGURE 9.3. Exponential decay of the price of the Model-T Ford automobiles during the period from 1909 to 1923. The prices are in constant dollars. P is the price in any given year, P_o is the initial price of $3,300 in 1909, and P_s is the saturation price of $800 that would have been reached if the model had not been discontinued. Source: W. J. Abernathy and K. Wayne, "Limits of the Learning Curve," Harvard Bus. Rev., vol. 52, 109 (1974) p. 109.

obtained by assuming that the stable price would have been $800 if the model had not been changed. The slope of this straight line yields the characteristic decay time of 4.5 years.

More modern data indicate that the price-decay law presented here is rather universal because they represent a different period of time as well as a quite different product—namely integrated electronic circuits. These were first produced in 1964 at an average unit price of $16. By 1972 the price had declined to $0.60 per unit (constant dollars). If a stable price level of $0.45 per unit is assumed, the data for integrated circuits follow the exponential decay line shown in Figure 9.4. The characteristic decay time given by the slope of this line is 1.8 years; a very fast decline indeed!

The price-decline behavior described here is mathematically equivalent to the behavior described by "learning curves." Readers who are interested in a proof of this equivalence are referred to Gilman.[6] However, the price-decay law given here is more convenient for theoretical studies and for applications because it is consistent with the many other aspects of business problems that are time dependent. Some of these are interest charges, sales growth, earnings growth, and cash flow.

Underlying both learning-curve theory and price-decline theory is the empirical fact that manufacturing-costs tend to decline exponentially with time.

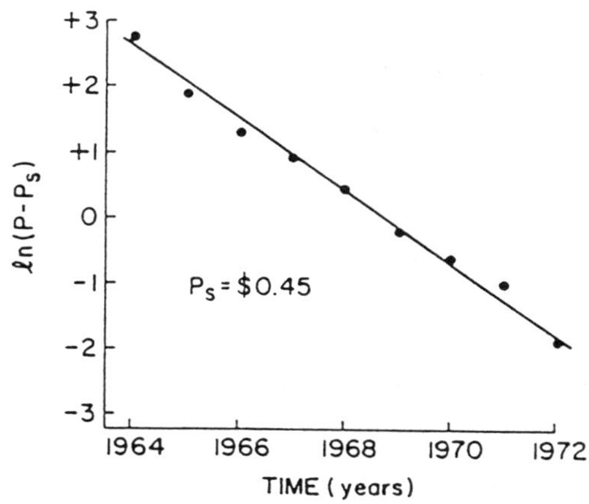

FIGURE 9.4. Exponential decay of the prices of integrated electronic circuits during the period from 1964 to 1972. The prices are in constant dollars and taken from data of the Electronics Industry Association. A saturation price of \$0.45 gives the best data fit.

This aids in predicting future costs, in the development of plans to reduce costs, and in monitoring the time history of costs for comparison with plans.

DIFFUSION THEORY

For estimating the costs, revenues, and financial returns during the diffusion of a new product or technology, more than the form of the kinetic process is needed. One must also have estimates of the coefficients; at least the saturation size of the market, and the initial penetration rate are required. Market shares, and any rate changes that may occur, are also very desirable as well as cost and price-change information. Methods for estimating these quantities have been proposed by various authors with one of the best models being that of Mansfield-Blackman.[7] They determined some of the factors that determine the driving force for introduction of a new product, and therefore the market penetration rate. These factors were defined as follows:

I = Innovation index for particular industrial sectors
P = Profitability index = innovation's rate of return/ minimum expected ROR
S = Investment index = cost to purchaser/ assets of purchaser

A statistical study indicated that the rate of penetration of a new technology could either increase or decrease with I, would increase with P, and would decrease with S. The most innovative sectors were found to be the aircraft, electrical (including communications), and chemicals sectors. Their indices were 2.29, 1.76, and 0.60, respectively. The least innovative sectors were the paper, textile, and rubber sectors with the indices −0.75, −0.75, and −0.76, respectively.

Within a particular industrial sector (as defined by the U.S. Bureau of Labor Statistics, for example) the rate of penetration increases with profitability, and decreases with the relative size of the investment (for example, an individual changes shampoos more quickly than automobiles unless the person is very wealthy).

Diffusion of technology in both time and space has been discussed by various authors. See Mahajan and Peterson, for example.[8]

NOTES

1. J. C. Fisher and R. H. Pry, "A Simple Substitution Model of Technological Change," *Tech. Forecast. Soc. Change*, Vol. 3, (1971) p. 75.
2. J. J. Gilman, "Market Penetration Rates and Their Effect on Value," *Research Management*, Vol. 25 No. 2 (1982) p. 34.
3. Fisher and Pry, "A Simple Substitution Model of Technological Change."
4. E. W. Montroll and W. W. Badger, *Introduction to Quantitative Aspects of Social Phenomena* (New York: Gordon and Breach, 1974) pp. 19–39.
5. A. P. Hurter and A. H. Rubenstein, "Market Penetration by New Innovations: The Technological Literature," *Tech. Forecasting and Social Change*, Vol. 11 (1978) p. 197.
6. Gilman, "Market Penetration Rates and Their Effection Value."
7. A. W. Blackman, "The Market Dynamics of Technological Substituions," *Tech. Forecast. Soc. Change*, Vol. 6 (1974) p. 41.
8. V. Mahajan and R. A. Peterson, "Integrating Time and Space in Technological Substitution Models," *Tech. Forecast. Soc. Change*, Vol. 14 (1979) p. 231.

10

FINANCE AS IT EFFECTS RESEARCH POLICY

Both the entrepreneur and the financial backer need the right environment: low interest rates, low cost of capital, low inflation, low taxes on long term capital gains, and high savings rates. No laws of economics need to be broken, no one's life style needs to be crimped to realize these conditions. Both Japan and Germany have achieved them all.

—Simon Ramo, *Scientific American, 260*, 148 (1989).

To those who were active in industrial research through the period of the Vietnam War (circa 1970) a large change occurred in the attitudes and the resulting policies of managers in the United States. Related changes occurred in management policies for university, government, and nonprofit research organizations later.

An attempt to capture the nature of this discontinuity will be made here by using the interview style again with Mr. George Kazan. He experienced it as a high-level manager in the Allied Chemical Corporation (now Allied-Signal).

KAZAN: Somewhere around 1970 there was a big change. Financial and legal people began to exert the upper hand in corporate management. They had control over the boards of directors who selected the chief operating executives, the chairmen, and the presidents. How did this come about? In the companies I knew, it came about because there was a failure to methodically train people for the top levels from within the companies. To have a planned succession, such as that in a company like Amoco, a Fuller and a Morrow were trained by Swearingen (Morrow succeeded Swearingen, and Fuller succeeded Morrow as the CEOs of Amoco). This did not occur in the companies that I have observed closely such as Allied Chemical, Olin, W.R. Grace, and B.F. Goodrich. It did occur at DuPont where there is a succession planning ritual.

When dissatisfaction with a chairman occurs, it leads to a situation in which outsiders: financial people, bankers, investment houses, and lawyers manipulate the board of directors via the dissatisfaction with the performance of the company. They use the latter as a lever. "That bad man you've got in there—the chairman, or that bad president—has to be replaced. We can't put up with him. We can't let our good money be spent by a fool." They recruit a replacement. And in companies with plenty of internal candidates, they recruit within the company. In companies with a scarcity of candidates, or an enhanced dissatisfaction, they recruit outside the company. When they recruit outside the company and install a new person in a B.F. Goodrich, a W.R. Grace, an Allied, or an Olin, the people who install the replacement have the power. There is an obligation from the newly installed chairman, or the new president.

The mentors of the new person are the financial people on the board who did the selection and recruitment. They have the power to replace the person. They have the power to control his or her compensation, bonus, pension, and any other extraordinary incentives and rewards. At Allied Chemical, until 1967 when the board became dissatisfied with Chet Brown and replaced him with Jack Connor, I never knew a situation where the chairman was a stranger.

The game played by the new, outside chairman is likely to be getting the board to pay more, give bigger bonuses, and give a never-deny-my-person-anything kind of approval. The beginning of disapproval is when they deny approval of a capital authorization request, for example. When a person is in favor and the board thinks he or she is doing well, everything asked for is given. When the board becomes dissatisfied, the person knows there is trouble when denials begin. Every proposition made by the chairman is presold to the board members when they approve, so it is even more telling when a negative comes out because he or she has already tried to presell the proposition. The chairman half knows then that it's not going to be accepted. When it's formally presented and rejected, the chairman knows for sure that his or her position is in trouble. Thus the power is given to people who are not professional creators of new businesses.

The DuPont chairman, Jefferson, won board approval of the Conoco acquisition. This made them put up 7.7 billion dollars, all debt financing, to get into an oil business that was at the mercy of variations in the selling prices for crude oil, gasoline, oil products, refined products, and chemicals. It quickly became clear that Jefferson had chosen a mediocre oil company, in a very competitive business. He also told the press that he was going to put the company into the biotechnology business. Announcements claimed that they could synthesize industrial organic chemicals, engineer new plants that could fix their own nitrogen, invent new plants for large-volume agriculture, and prepare new and old pharmaceuticals; all by means of biotechnology.

GILMAN: It soon became obvious that they couldn't quickly exploit biotechnology.

KAZAN: Yes, then Jefferson reached retirement age and was replaced by R. E. Heckert who retired in turn after a few years. From an outsiders point of view the faith in biotechnology has not been justified.

This disease, manifest by dissatisfaction with top people, started about 1965. Boards of directors started replacing chairmen and presidents inside and outside of companies, hunting for leadership. In a vacuum of unsatisfactory leadership the people who make the decisions to replace, or not to replace, have the real power. The fact that they can control the employed-period income, and the post–employed-period income is the power. The individual is at the mercy of that power. The chairman's incentive is to retire with as much money as possible. That becomes his or her aim.

The chairman's aim before I was 50 years old was to retire proud of technical prowess. You knew this was so by the fact that they couldn't stop talking about what happened to their technical offspring after they retired. They still had a consuming major interest in the company's product and process welfare when they were no longer boss.

The financial types, the lawyers, the bankers, the persons who would not be so proud of technology, cared far less about what happened to the company after they left. For example, even Jefferson (a chemist) said, "when I get through or when I'm retired, I'm not going to worry about the DuPont Company. I won't care about anybody's salary, or anybody's compensation. I'm not going to be uptight about anything I used to be uptight about." People 20 years before Jefferson never felt that way. They retired with regret, they tried to postpone retirement.

They loved what they were doing, if I can use that word. They were deeply turned on by the activity, by running the giant machine, and by creating things that made the giant machine bigger and better.

At Allied, before 1965, it was said, "you never make any money by going out of business." That's debatable. Some people think that when businesses are bad you may become better off by going out of the business. But such decisions are not always right.

For example, at Allied, when the decision to get out of the sulfuric acid business was made, they had something like 25 sulfuric acid plants covering the whole country, making 7 or 8 thousand tons a day of acid, and making 11 or 12 million dollars a year of operating income. The decision to quit the business was questionable. It was based on a flawed analysis. On the other hand, the decision to get out of the dye business was probably correct because there it was a clear loss of competitiveness. They could no longer compete. Prices had declined so badly as a result of severe competition that you could be sure it was best to quit. For sulfuric acid, that wasn't the case.

At the time period under discussion, the competition for technology became severe; the cost of money rose to new heights; and the selling prices of products were hurt by competition. This led to reexamination of ownership by a great variety of companies. It led, for example, to companies like DuPont getting out of the dye business. Dye stuffs had been one of their proudest achievements of the post–World War I period.

GILMAN: Another change was that, because of competitive factors, the cost of invention went up at a very high rate. Competition for people increased. In part because of the expansion of military R&D. Also because of the expansion of university departments. Therefore, the cost of getting good people to work in industrial research laboratories went way up. This was good for the technical community from the point of view of their compensation, but it contributed to reducing the rate of research investment on the part of the corporations.

KAZAN: Let's try to put some numbers on that. I was hired as a bachelor in chemical engineering in 1939 for $1,500 a year. The cost of maintaining me was about $4,000–5,000 a year. People with doctorates in chemistry at that time were hired at $2,700–3,000 a year. The cost for engineering and physical sciences graduates in 1939 was an average of about $5,000 a year. The cost for doing just that in 1972 was about $100,000 a year. The cost at DuPont today is about $200,000 a yr. In the oil companies it may be $250,000 a year per professional, or more.

So there has been a giant increase in the minimal expense of having inventive people in place.

GILMAN: Another financial factor that has changed is the real interest rate; that is, the nominal interest rate minus the inflation rate. The real interest rate is currently in the neighborhood of 5 or 6 percent, whereas back in 1965–70 it was of the order of 2–3 percent. So the real interest rate has roughly doubled. This means that the prudent patience time (the time that one should be willing to wait for an investment to begin to return earnings) has changed by a factor of e^2, or about 7.4, which is a large factor indeed. It seems to me that this large reduction in the length of time that people are willing to wait is what underlies many current research management problems.

For our competitors the situation has been quite different. Japan's real interest rate was under 2 percent for a long time. Now it may have risen. Until recently, European rates have tended to lie midway between those of Japan and the United States.

KAZAN: I come from a tradition where a business is a prize. They are held as prizes, they are good, they are not bad. When you have a prized child that you brought up and it develops a flaw, let's say metaphorically it's unhealthy, then you try to get a doctor for it. When you have a business that you've always prized, that's always earned it's keep and rewarded you well, but it no longer does so—that normally isn't the time to call for the undertaker—it is the time to call for the physician. I am sure that this is apt for the technological industries that we're concerned with.

The purpose of having a large staff of people is to let them solve the problems of this ailing child, this thing that isn't what it used to be. Whatever these problems may be. It may be that the disease is mortal and that changes everything. Although I know of cases where it's mortal, and needs to be quietly buried, I've known lots of cases where people want to call it mortal for reasons that were specious. Reasons used in an analysis that says "this business is ready to be disposed of and buried" are frequently flawed. You may have to work hard to find the flaw, and you may have to bring to bear some competence that relates to the abilities of your problem solvers to recommend a no-ditch procedure.

Sulfuric acid technology, to use a very old one, was throughout my lifetime constantly improved. And it seemed like the oldest, dullest thing you could do, but today people are still making a lot of money from it. The real question is: does it earn more money than the real interest rate? And: can it withstand variations in the real interest rate? One of the features of the real interest rate is that it varies with time.

Because of the time dependence of the interest rate, you need to demonstrate your analysis of operating policies for the worst case, the case of highest real interest rate. You need to look at it with a real interest rate projection that takes into account the economic factors that make real interest rates cycle. You can't take a point condition and say: here, June 7th, 1989, the real interest rate in this country is 4 percent and therefore since I'm only earning 3.5 percent now, I ditch. You really should be brave enough to say this business will take the ups and downs of this factor. If you could prove to yourself that real interest rates are going to increase permanently, it changes the whole analysis.

GILMAN: Yes, that's right. It also changes what is a successful invention compared with what is not a successful one. Within it's own context, of course. You can take the same invention and move it to a different country where the financial factors are different and it may change from poor to good.

KAZAN: So you have to consider the value of this package, inside and outside your own country, inside and outside your own culture. A great example is Chairman Hennesey's decision at Allied to spin Henley off. He kept everything he thought was a prize by the analysis of himself and his co-workers. But Michael Dingman showed that many of the things that were ditched were very valuable, so one of the things that analysis begs for is examples of the flaws in behavior.

One of the things that's important in considering inventivity is the host environment. There are some environments that are quite capable of generating specious reasoning and flawed proofs. No amount of inventivity, or money that goes into establishing inventive staffs, will put the necessary resources together in a bad host, it won't work. One of the things about people who wish to put a lot of money into the establishment or embellishment of an inventive staff is the host itself. There are managements that want an inventive staff for bad reasons, that will let research funds go into a bad host. Such a management has another purpose in mind. But such situations are outside the scope of this book. It's not a book on ethics. Right?

GILMAN: Right.

KAZAN: The assumption in this book is that if people sincerely want to bring inventions into the world that allow new products to be made by creative people, the return on investment for these products is at least a substantial amount over the real interest rate. The question

> becomes, are there any children who are undesirable. There are some products that return quadruple the real interest rate, and that's great. But there are some that will potentially return 50 percent over the real interest rate; they need to be questioned. The pursuit of them has to be weighed carefully with the risks. There are lots of intangibles in any decision. My history is that new products coming out of good inventions always go into the world sailing way over real interest rates.

My interpretation of what happened around 1970 is that a quantitative change in the financial system of the United States occurred. This was a direct result of the Vietnam War having been financed through debt instead of current tax revenues. It led to a seller's market for money combined with inflation. Because of the former, the spread between the rate of inflation and the cost of money increased substantially. Because of the latter, the absolute cost of money (the absolute interest rate) increased discontinuously by a factor of about 2.5. These effects of the war financing were further aggravated (or stabilized) by the discontinuity in the price of oil that was induced by the OPEC cartel.

Secondary effects of the financial changes intensified the situation. Since growing one's own real assets is a slower process than buying them, companies had an increasing tendency to buy other companies; thereby acquiring assets. But the buying was done not with cash but with debt. Since debt is the equivalent of money, this increased the supply of money, thereby contributing to further inflation. It also increased the demand for money (read debt), thereby pushing the price up (that is, the real interest rate). Thus stagflation occurred.

Financial behavior is an abstract representation of social behavior. They are functions of one another. Therefore, it is not surprising that attitudes changed as the financial parameters changed. In other words, the financial changes caused the centers of power to shift. And caused the focus to shift from deferred income to current income. Also, the heights of the hurdles for the expected returns on investment rose.

These factors caused research managers and their superiors to change their attitudes toward the research enterprise. First in industry, then in government, and finally in universities. The change was to a short-term outlook regarding research. Not because this was natural for them, but because there was no other realistic response to the changes in the financial environment.

Direct financial factors were exacerbated by less direct ones including: tax policy, regulation, and antitrust policy.[1] High capital gains taxes combined with double taxation of dividends discouraged emphasis on real asset generation. Regulation syphoned funds from long-term investments into short-term "fixes." Antitrust policies led from homogeneous companies with technical depth, to technically shallow conglomerates. All of these policies led to financially oriented managements with short-term horizons.

A result of all this is that no number of researchers talking to each other, or talking at research managers, or criticizing business managers, is going to change the current short-term focus. If a change is to occur, the behavior of the financial system through its quantitative parameters must be changed. But no one has socially acceptable ideas for doing this.

Like any foolish action, financing a war by means of debt is not a winning long-term strategy and is very difficult to reverse. The Romans learned this, as did the Spanish, French, British, Germans, Russians, and others in times past. Now the United States is struggling with it. Only the Japanese and Germans seem to have learned how to reverse the process.

NOTES

1. P. F. Drucker, "Science and Industry, Challenges of Antagonistic Interdependence," *Science*, Vol. 204 (1979) p. 806.

11

TRANSFORMING INVENTIONS INTO COMMERCIAL BUSINESSES

For a successful technology, reality must take precedence over public relations, for Nature cannot be fooled.

—R. P. Feynman, *What Do You Care What Other People Think?* p.237, W. W. Norton, New York, (1988).

Many steps, much time,. plentiful money, and multitudes of people are needed to transform inventions into viable processes, devices, or products. This should be clear from the preceding chapters. Even seminal inventions like the oft cited transistor didn't jump from its womb and dazzle the world. I first learned about it when J. Bardeen presented a physics seminar at Columbia University in about 1950. At that time he spoke of it as an interesting device, but was apologetic about its commercial potential because it was somewhat noisy, and had a limited frequency response. He demonstrated its meager capability in an audio amplifier. For its first few years there were also serious stability problems for germanium-based transistors because the germanium was sensitive to atmospheric moisture. Thus, in their early lives transistors seemed more like curiosities than something that would lead one of the most profound social changes in the history of the world. Silicon came to the rescue.

Although patented inventions and their subsequent development are the best way to get new businesses started (because of the temporary protection they provide from competitors), they are not adequate of themselves. And they are not necessarily the best way to improve an old business—or to diversify. They may, in fact, lead to the "diworseifications" described by Lynch.[1]

Since invention is just the first step toward a new business, a company that finances research with this purpose in mind should be prepared to persevere through many additional steps, and do so with some enthusiasm. It is unrealistic to expect that the inventor can do it himself. It is hard, often discouraging work, so it is unlikely to be done well without solid support.

The initial invention is usually rudimentary. It must be developed into a prototype through further research or through engineering design work. Then it must be produced in small quantities (or if it is a process it must be used to produce small quantities). The results must be presented to potential customers who almost surely will want modifications so some redesign must occur. There may be more than one cycle of this. Then an initial manufacturing facility must be built and operated, marketing methods must be established, and an administration must be organized.

COSTS OF DEVELOPMENT AND COMMERCIALIZATION

For some unfathomable reason general managements seem to think that the staff that creates an invention should be able to develop it into a product and commercialize it with little augmentation of their existing budgets for operating expenses and capital expenditures. Or they think that the invention can be put into an existing business operation and then be developed and commercialized using the operating margin of that business. Both ideas are silly in the light of well-established and known operating ratios that cannot be escaped, not to mention the poor motivations of the research staff and business managers under these conditions.

Consider the ratios just mentioned. One of them is the investment/sales ratio. Long experience teaches that in order to obtain one dollar of sales, a manufacturing company must invest approximately one dollar of capital plus or minus a factor of two. For an operation of low capital intensity the investment might be 50 cents, while for a high capital intensity one it might be two dollars, but this approximate range cannot be escaped. Trying to get away with underinvesting simply means that the investment will be wasted, which is hardly good husbandry.

Further evidence of the validity and prudence of this ratio is provided by the stock market. Its role there has been discussed at length by Fisher.[2]

Another ratio that can easily be obtained from sources of corporate statistics is that the sales revenues of manufacturing companies fall in a certain range on a per employee basis. This varies with the real value of money. Therefore, a better ratio is that of revenues to the average employee wage, but wages are not ordinarily published so this ratio is not easily determined. Again this ratio falls within a limited range. It means that commercialization requires staffing at a level that is consistent with the revenue expectation. If it is attempted with significantly fewer people it will probably fail, and again whatever has been invested in it will be wasted.

Initially there is no good way of knowing whether the capital and operating costs of development and commercialization are "expense," or "investment." Thus it is not surprising that the managers of ongoing businesses do not embrace them with enthusiasm. They mean that transforming inventions of research into real products is expensive and can only be justified by the existence of significant markets.

TRANSFER OF TECHNOLOGY FROM GOVERNMENT LABORATORIES

While the business world tends to underspend when trying to transform inventions into commercial products, the government world tends to overspend. In some cases a swamp of unproductive money is created before the invention has even been made (as discussed in Chapter 8). In other cases inventions are mortgaged with such high levels of initial expenses that they can never create positive cash flow (over the complete life of the activity). Then they either sink, or require subsidization in perpetuity. A good example is the situation with solar energy. There are quite useful small scale devices powered by solar energy. These comments do not apply to them. However, the main thrust of the government's program was toward solar power stations, and in the 1970s billions of dollars were spent on it. A research agency was set up, including a research institute (Solar Energy Research Institute) that continues to exist twenty years later (and perhaps two or more billion dollars later). Yet no unsubsidized power stations exist. If there ever are any, they will start off with cash-flow deficits of hundreds of millions of dollars (the future value of the money spent 15–20 years ago).

Other examples abound: energy from nuclear fusion, breeder reactors, some geothermal projects, composite materials, ceramic gas turbines, railways levitated by superconductors, gasohol, and many others.

In part the problem is a matter of timing. That is, pushing a technology before the conditions, or the techniques, are ready for it. But in large part the problems are caused by ignoring, or suppressing, the flaws of a technology. Thus solar power is flawed by that inescapable fact that the power *flux* of sunlight incident upon the earth is very low (by several orders of magnitude) compared with the power flux that is needed to operate most devices in factories and homes. Therefore, expensive machinery is necessary to concentrate it up to the needed levels.

The other listed technologies also have flaws. Otherwise, commercial interests would put them to use without government intervention and/or subsidies. One further example is the case of ceramic gas turbines. At first sight it is a fine idea, but there are two flaws. Ceramics are intrinsically brittle and expensive to fabricate. After all, if brittleness were not a serious problem, we would still be building wooden airplanes.

In general the problem with transferring technology from government laboratories is the lack, in them, of serious motivation to consider inventions as parts of systems, including the economic aspects of the systems. Short-term political power drives their activities, but in society as a whole economics wins in the long term.

Dilemmas of New Ventures

Transforming inventions into commercial activities usually means that a parent organization must find some way to give birth to a daughter organization. Furthermore, if the invention is truly novel, the daughter will bear little resemblance to the parent. These conditions lead to dilemmas, not to straightforward procedures.

The first dilemma: how can it be determined that an invention is likely to lead to sustained commercial growth if the company in which the invention occurs is not already in the business? Note that important inventions are likely to be of this type. That is why they are inventions rather than improvements. A closely related dilemma: how can a company become sufficiently aware of whatever the competition is to make an assessment if it is not tied into the technical community that surrounds a particular invention? All too many companies have bought licenses to practice particular patents where the evidence of effectiveness was only the word of the inventor and the impression of the patent examiner.

If an invention is defined explicitly, it is relatively easy to test whether it can be reduced to practice. But most are not well-defined, and even if one is, there still remains the question of how it fits into the marketplace.

Some companies try to solve this first dilemma by taking a minority interest in a company that sells to the marketplace of interest. The rationale is that this will give them a "window" to the business area, and if they like what they see they can buy a majority position later. But this leads to further dilemmas. One is that they don't really get a clear view of the business. As a minority owner, their participation is vicarious. They may get a seat on the board of directors, or two, but the board only gets to know what is presented to it by the management. The outside board members do not participate in day-to-day activities at all.

Another problem of taking minority positions is that whatever resources are put into the external company get diluted. For example, suppose that a company owns 20 percent of an external company, and then puts a new product into it as a means of exploiting the product. Then its potential earnings from the product are diluted 4:1. But that's not all! Suppose the new product is very successful. So much so that it doubles the earnings of the external company, thereby increasing its P/E ratio by 50 percent. The cost of buying further equity in the company will have increased by a factor of three! In effect, the cash of one company will have been used to create prosperity in another one. This is not likely to be considered prudent by the first company's stockholders.

If the external company is small with relatively few public stockholders, purchasing a substantial minority of it privately (say 20 percent) is treacherous even as a pure investment, particularly if the purchaser is a much larger company. Such an investment may be difficult to liquidate. First, because the thin market makes the price volatile. Second, because sales of private shares will undermine confidence in the company. And third, because both of the previous points are likely to infuriate the other stockholders in the small company. This may lead to litigation.

If, for the reasons just described, minority equity purchases do not make attractive external ventures, it might be thought that straight "buy-outs" would be better. But it is well-known that such buy-outs almost always lead to severe personnel problems in the bought-out firm. This creates a third dilemma. To the extent that the buy-out makes people in the external firm financially secure (that is, rich), it also makes them psychologically independent (that is, prone to quit). Also, ownership produces an irresistible urge in the owner to treat the owned according to the golden rule (that is, he who has the gold makes the rules). This decimates the morale in a proud small company. The time needed for it to happen is about the same as the time it takes for an auditor to open his notebook.

An operating mode that bypasses many of the problems of straight buy-outs, and minority equity positions, is that of joint venturing. In this case a jointly held company is set up in which the external firm has minority ownership. Thus the initiating company has operating control. Typically, the initiating company invests cash in the joint venture, while the external company invests some of its stock plus some technology, or other expertise. A small external company is usually cash short, so it should not be expected to contribute cash to the joint venture, but a stock contribution from it is very important because it generates a serious interest in the success of the joint venture.

As a mode of operation, joint ventures do not appear to have fundamental flaws, but their success depends on satisfying some constraints if the two companies that are involved are disparate in size or have distinctly different levels of expertise in the business area of interest.

For example, consider a venture between two companies: Huge and Tiny. Tiny's interest in the venture will probably be based on a need for money, while the interest of Huge will probably be to increase its long-term return on equity by entering a business that is growing more rapidly than itself.

The natural tendency of Huge will be to constrain the joint venture, thereby reducing its growth rate below that of Tiny (which was what attracted Huge in the first place). In the meantime, Tiny may well be growing rapidly in its other activities. It may be able to fund these through internal cash flow, or other sources. Under such circumstances, Tiny will soon lose interest in participating in the joint venture. As the ratio of its equity in the joint venture to its total equity decreases, so will its interest in operating the venture. But Tiny's technical expertise is probably crucial to the acceptance of the venture's product in the marketplace, so this lack of interest will have devastating effects.

As a result, it is crucial for Huge to manage the joint venture as a decentralized subsidiary in order not to destroy local morale or decrease rapid decision making. It is also important for Huge to ensure that the joint venture is not dependent on Tiny for technical knowledge, manufacturing, or marketing. Otherwise it will be vulnerable to the "empty shell syndrome." All three companies should cooperate with one another, of course, but the joint venture should be able to operate nearly independently.

It may be seen that a joint venture may easily become a dilemma if it is not managed skillfully. But in this case dilemmas can be avoided if those involved are sensitive to the fact that they must carry out a delicate balancing act. As a successful joint venture matures, the problems previously outlined will tend to disappear, so the discussion has been directed at the early stages in the development of a venture.

A final operating mode that may sometimes be appropriate is that of a staged buy-out. I once got involved in this mode unintentionally as a means for solving a problem. During the process, I learned that it often has some virtues.

In an area of rapid technical growth (excimer lasers), I placed a contract for the design and construction of a piece of hardware that could be sold as a product if it met the design specifications. The contract was with a small start-up company, and the plan was for us to work together on applications of the hardware, as well as on its marketing.

The design phase was quite successful, but as a result of the rapid technical changes in the area, the marketplace had changed by the time the hardware had been developed (about two years). Therefore, my company did not want to try to market the prototype. We also did not want to scratch the project, since we already had about 1 million dollars invested in it. In other words, we faced a very concrete dilemma. For the reasons I have already discussed, we did not want to take a minority position in the small company that had developed the prototype, nor did we want to buy it outright. We considered setting up a conventional joint venture, but did not think that we had enough people on our side with the necessary technical skills to run such a joint venture successfully. In brief, we had worked ourselves into a tight corner.

Our solution to the dilemma was to sell our technology (the prototype and its design) back to the company that had done the design and construction. Not for cash, because they didn't have any. But for stock. We also bought additional stock for cash in order to give us a substantial, but not a majority, interest. Furthermore, we agreed to repeat this process for a few years until we reached full ownership. That is, we agreed to place hardware development contracts with the small company for items that we selected. Then, given a successful piece of hardware, we would sell it back to them for more stock at predetermined prices.

This scheme was designed to keep us involved in the technology during a period when our fractional ownership was increasing. Also, during this period Huge could not smother Tiny because it would not have control. And the owners of Tiny were kept involved because they would not get a large payoff until the end of the process, and the size of the payoff was designed to depend on the progress of the business in the meantime.

Staged buy-outs are not a universal solution for the dilemmas of new ventures, nor are they easily negotiated, but they may help in solving some of the problems.

NOTES

1. P. Lynch, *One Up on Wall Street* (New York: Simon and Schuster, 1989).
2. K. L. Fisher, *Superstocks*, (Homewood, IL: Dow Jones-Irwin).

12

IMPROVING THE PERFORMANCE OF RESEARCHERS

In those great two years he (Newton) had to the full two priceless gifts which no one enjoys today, full leisure and quiet. Leisure and quiet do not produce a Newton, but without them even a Newton is unlikely to bring to ripeness the fruits of his genius.

—E. N. daC. Andrade,
in Newton Tercentenary Celebrations,
Royal Society of London (1947).

During the period that Andrade refers to above, I doubt that there were complaints about Newton's performance of his research activities. But who was setting his objectives? Himself, of course. Who else could have? Who was measuring his performance? He was. Only later could his contemporaries make judgments. He was out in front. They were in his wake.

Newtons are very rare, of course. Their performance represents an ideal pinnacle. But the same principles apply to lesser people. If they are doing inventive research, their objectives are internally determined. And even if they weren't, no one else would be in a position to judge them until after an invention had actually been made. This is why objectives that come from outside of inventors are unlikely to improve their performance. It is pure conceit on the part of a management to think otherwise. Is this widely understood? Apparently not, if various kinds of management behaviors are observed.

On the basis of these two paragraphs, one should not conclude that nothing can be done about improving the performance of inventive research workers. However, these paragraphs do suggest that what is done should be centered on facilitating, rather than subjugating.

This issue is central to research management, so I will try to say it in yet another way. It is not that management *should not* determine the detailed direction of the inventor. It is that management *cannot* determine it, simply because the direction does not exist until the magic moment of conception or of discovery. Since management has been appointed to "command" the workers, this situation is intensely frustrating to it, and it is easy to understand why. However, if it succumbs, and all weak managements do, the magic moments are unlikely to ever arrive.

Problem solving in development work and engineering design is different. For it management needs to be very much involved in defining the problem, helping to find solutions, and determining when a solution is satisfactory. A skilled sophist will immediately point out that the line of demarkation between inventing and problem solving can be fuzzy. However, it need not be if one remembers that inventing has a legal definition. The output of inventing is patentable entities, and whether an entity is patentable is determinable by means of a hierarchy of official judges beginning with a patent examiner. The key word here is patentable since not all inventions are actually patented. But if an entity is not patentable, it may be the solution of a problem, but it is not an invention.

It may be concluded that direct approaches to improving performance (increasing inventivity) are likely to be counterproductive. This leaves a variety of indirect methods to be considered. They will not be discussed in any particular order, and some are more likely to be effective than others. Also, some will be more subtle than others.

The best opportunity for improving the performance of an organization, by far, occurs each time someone new is hired. Therefore, the most important question for management to ask itself at that time is, will this person raise the average level of performance in this organization? Will the candidate bring new skills and/or knowledge that will increase our existing pool? On balance, will the candidate teach us more than we will teach the candidate? There is more discussion of this in the section on recruiting.

Although it may seem harsh, the next best opportunity occurs when someone leaves. Either through normal attrition, through finding a better situation elsewhere, or through forced attrition. If the organization is well managed, departees will, on average, be those who are less effective than the employees who stay behind, thereby raising the average. Also, hiring situations will be created giving opportunities for further raising the average performance level. Finally, the departee will often move to a more compatible position elsewhere and thereby become more effective, as well as more personally satisfied.

Notice that these comments refer to inputs to the research process—not outputs. Most organizations place entirely too much emphasis on outputs when they try to improve performance. They concentrate on "objectives." They should concentrate on strategies. These are the processes that get a process from some initial position (state) to some final position (state). The initial state is known, of course. In development work, the final state is also known, at least to a good approximation. What remains is the transitional process for getting from one to the other. In research work, the final state is not known. So it should be obvious that what needs consideration is the strategy. Instead, the "research objectives" are endlessly discussed. Poor performance inevitably follows.

Albert Szent-Gyorgyi, the biochemist, said it with passion:[1]

> *One factor which has filled my whole scientific life with agony has been writing project proposals. It seems logical that if one asks money from anyone with which to do research, one has to tell what one wants to do with it. However, the situation is not this simple because research means going out into the unknown with the hope of finding something new to bring home. If you know in advance what you are going to do, or even to find there, then it is not research at all: then it is only a kind of honorable occupation.*

One of the reasons that concentration on strategy is not a popular method for managing research is that it puts the burden for developing a strategy on the management rather than researchers. It requires that performance be improved through active facilitation of the work. Active facilitation may otherwise be spelled, HARD WORK. Not a recipe for popularity! Far more difficult and nerve-racking than simply sitting in judgment from time to time.

The first step in the facilitating of the efforts of a researcher is to assume competence, until proven otherwise. If someone does prove to be less than competent at doing research, there is no choice but to find something else for that person to do. If he or she is competent, or still assumed to be so, then by definition it must be assumed that he or she is honest. It follows that typical researchers can be given various powers that they can exercise at their own discretion. The most important of these for facilitating the work is financial discretion. That is, the power to personally authorize the expenditure of money in amounts up to some nontrivial level; say 2 to 3 percent of salary for a single requisition. Although this is seldom practiced, it is good policy simply because it saves expensive research time, and because nothing does more to encourage a sense of responsibility and worthiness in the researcher. Some will abuse such a policy. But this provides one way to find out who is competent and who is not.

The case was well put by Henry L. Stimson, who was the U.S. Secretary of War under Presidents Taft, Roosevelt, and Truman[2]: *"The chief lesson I have learned in a long life is that the only way to make a man trustworthy is to trust him."* This is an expression of the principle of reciprocal balancing in relationships between people. Friendliness tends to be met with friendliness, anger with anger, trust with honesty, and so on.

"Walking around" is another means for facilitating the work of researchers. If a manager has a question that a researcher with some particular expertise may be able to answer, or is curious about what someone is doing, a walk out into the shop is far more effective than a telephone call or requesting that the researcher come to the manager's office. Walking out to the researcher's office or laboratory accomplishes many things at once, some of which may be more important than the ostensible purpose of the walk. First, it "shows the flag" to several people along the path of the walk. Second, it honors and shows appreciation for the individual being visited. Third, it respects and may flatter the expertise of the individual. Fourth, it transfers primary information if the manager gets questions answered or curiosity quenched. Fifth, it transfers secondary information from observations along the path and allows the manager to "schmooze" for a few minutes. Sixth, it differentiates a manager who does it from the rest.

Researchers never forget a manager who takes the trouble to walk out to them. I know this from comments made to me many years after some of my walks.

I should warn that casual walking around is probably not a good idea. It is too easily interpreted as snooping, or spying. There should be a spur-of-the-moment legitimate reason for the walk. On the other hand it should not be a formal affair that is scheduled hours, or days, in advance. The main effect of such formal visits is to disrupt whatever work is going on.

If the organizational unit is too large, or for some other reason walking around is impractical, management still needs some mechanism for explicitly expressing interest in the work. And frequently. Morale is devastated if researchers get a sense that management doesn't really care about their output. That they are being supported for some political reason rather than the technical value of their work. One technique is an award system; especially if it deals in small as well as large awards. Just as important as the awards themselves is the fact that the management takes the time and effort to make them. Another is a technical newsletter. Still another is a periodic luncheon at which there is a brief technical discussion.

Attention to the details of day-to-day operations is of the utmost importance. Activities such as purchasing, library services, machine shops, electronic repairs, and computing are examples. Delays in providing these services should be minimized. Not for the convenience of the researchers, but to facilitate (reduce the cost of) the work. Research is a sequential process. A second step cannot proceed until a first one has been finished. A third step awaits a second. And so on. When distributed over a long sequence of steps, 10 percent delays at each step can easily accumulate to become a 100 percent delay overall. Thus speeding up operations in some cases will do more than anything else in improving overall performance.

Providing a framework may well increase inventivity. Although the most creative people construct their own frameworks, there is a larger, still very creative, group that feels uncomfortable in the absence of any guidance. There is, of course, a level of guidance that yields diminishing returns. A large amount of guidance leaves no room for invention. Knowing where the point of diminishing returns lies marks the difference between good and poor management. A framework that contains much unexplored territory will obviously be more helpful than one that was been worked over like the tailings of a gold mine. Thus management has a substantial responsibility to be met in setting up the framework. Experiences with some frameworks are discussed in Chapter 6.

REPORTING

Although they are resisted strongly, and sometimes bitterly, by researchers, reports written periodically can do much to improve productivity. They improve clarity of thought, find errors and omissions, and provide perspective. For most people these are unlikely to occur without some formal discipline. In order to write something that will be intelligible to someone else, an author must try to present an organized progression of ideas and facts. A random series will not do. Gaps in knowledge, logic, understanding, and perspective, are soon exposed by this process. A bonus is that the writers learn to express themselves.

It is not enough to ask the researchers alone to write periodic reports of their work. Each level of the organization should do it by writing for the level just above it. Thus a hierarchy of "diaries" will develop. Whether these should be daily, weekly, monthly, quarterly, etcetera will depend on the local situation. Periodicity is very important. It imposes discipline and rhythm (remember the musical analog). When I was managing a sizable research center, I imposed a weekly newsletter on myself, and an annual report. If there was nothing to report in a given week (and there may not be because of vacations, lulls, conventions, etcetera), I still issued a newsletter that simply stated that there was nothing to report.

Periodic reporting is good for morale in addition to several other virtues. Research work (and development work) on a day-to-day basis consists mostly of failures (remember Edison's 99 percent perspiration). Ideas turn out to be wrong, equipment breaks, supplies arrive late, and so on. Therefore, it is easy to get the feeling that there is no forward motion. But if there is a systematic record that can be reviewed from time to time, the progress being made becomes clear.

There is also a practical advantage of writing reports regularly at relatively short intervals—say weekly. It provides that grist for reporting to others who usually want reports on some more extended time scale. Suppose someone wants bimonthly reports. There is no need for a bimonthly scramble to gather information. One simply "cuts-and-pastes" eight or nine of the existing weekly newsletters. Even annual reports are mostly a matter of collating the materials that have been collected in small increments each week.

It is almost needless to say that weekly newsletters tend to generate competition within an organization. Everybody wants to be included. And, as the number of available items increases, the standards for inclusion increase. Over a period of a few months, or years, the competition becomes intense. And no one has found it necessary to berate the workers or crack a whip!

For individual research workers, weekly reports are too frequent, although they should be encouraged to keep running notebooks. On the other hand, it seems to me that they should be required to write annual reports consisting of a few thousand words. Such reports will force them to take an inventory of what they have done, to organize it into some kind of structure, and to contemplate where they are going. Such reports will also help them communicate their work to other people.

Oral reporting is probably even more important than written reporting. An author may hope that few, if any, will read his or her report; but a speaker has no place to hide. Feedback, be it positive or negative, occurs in real time. If it is positive, the speaker becomes exhilarated. If it is negative, depression results. In both cases the speakers will be powerfully motivated for the next occasion. The one who got positive feedback will want to feel more of the exhilaration experienced the first time. The less fortunate one will try hard to do better in order to avoid the feeling of depression. And, management need not do, or say, anything in the meantime, although help will be welcomed by both the successful, and the unsuccessful, speaker.

After a few repetitions of oral reports, a marked improvement in the quality of the reports will be apparent. Shy speakers will not become orators, but their reports will no longer ramble and will have acquired structure with a beginning, middle, and end. Most reports will be to the point, conveying specific information, instead of using all of the available time for an introduction, for example. The less experienced speakers will learn from the more experienced ones. Thus the exercise will become much improved, both by making the researchers better speakers and by conveying more information from one part of the organization to other parts.

INVENTION MEMORANDA

Another form of reporting is the disclosure of inventions. Most companies have an official format for doing this, but it often involves too much paperwork and therefore inhibits rather than enhances invention reporting. As has already been mentioned, at the time of their conception most inventions are very delicate entities, so delicate that most don't survive. Therefore, they need to be nurtured in all possible ways. Furthermore, the inventor needs to feel parental about the invention, but not defensive.

If the invention is put into the form of legalistic documents too soon, the inventor loses the sense of proprietorship, and starts to respond to the needs of the documentation (patent department) rather than the needs of the invention.

This can be alleviated through the use of informal "memoranda of invention" as precursors to formal invention disclosures. One page, bare bones, forms with space for a one or two paragraph description of the invention suffices. This will provide a record for establishing a date of the invention (it should be signed by a witness, or a supervisor), and a document for further discussion and consideration. Because of its informality, if the inventor decides not to pursue the invention, there is a minimum of embarrassment associated with "shelving" it. Most such memoranda will in fact be shelved—about 90 percent of them.

The memoranda that are pursued can be converted into formal disclosures and then patent applications, and two or more of them can be combined to form one or more cases. They are also convenient for review by managers and patent lawyers.

MAINTAINING STANDARDS

As just described, one method for setting, maintaining, and raising standards of performance over time is internal oral reporting. It is a short step to external oral reporting, and this is a most important reason why researchers should be encouraged to give oral presentations at various professional society meetings. By doing this they quickly learn what the standards of their profession are. Most will then strive to bring their own performance up to as high a benchmark as possible.

The same is true for written communications. Submitting a manuscript to a journal and receiving a response by a hostile reviewer is a baptism by fire, an unforgettable educational experience, and a test of an author's intent. It summarily separates the amateurs from the professionals. It is brutal, but it certainly sets a minimum standard for effective work.

Some research organizations have nonpublication policies because they think that they will lose valuable secrets if they publish. This is a possibility, but it is my observation that if they do not publish, their standards tend to sink so low that they not only have no secrets worthy of coveting, but their staffs become substandard. Sensible discretion should be used, of course, in carrying out the publication process.

Just as the commercial marketplace determines the values of goods and services, so participation in professional activities determine the values of research organizations. Self-evaluation, although widely practiced, is a delusion.

Internal provincialism is as bad as external. Researchers can gain understanding and solve problems more quickly if they use each other as consultants; thereby complementing one another. Management can bring them together in various ways. The most direct method is to put people who should benefit from interactions into adjacent laboratories, or offices. Or put them into the same organizational units. Study groups are also sometimes effective.

Seminar speakers invited from the outside not only bring in new information, but they also transfer standards from outside to inside. In addition they may bring solutions to specific problems, and they provide contact that may well be important for recruiting. Thus the notion of some managements that they represent costly "frills" is false.

RATINGS

"Systems" for rating performance are rampant. They are designed by consulting firms commissioned by personnel departments. Thus they tend to be vested through their expense (no one wants to admit that they aren't effective), and they are counterproductive for all but the convenience of the personnel department.

Why are they counterproductive? Because they tend to maintain the *status quo*, while people and conditions are changing. Remember it was said earlier (Chapter 4) that the most important part of personnel management is to match motivations and jobs. Rating systems tend to prevent this. The way this happens is as follows. Suppose someone is motivationally mismatched with a job. As a result, this someone performs in a less than excellent fashion (remember that motivational mismatches usually mean that the person's skills do not match the job). So, the person is given a mediocre rating. In a large organization (and often in a small one) the odds are good that some other job will provide a better match for the person. But this other job will not be offered to the person because of the unattractive performance rating.

There are two direct consequences of the phenomenon just described. One is that ratings are soon corrupted, and the other is that they are used punitively. Corruption enters because managers usually do not want to hurt people, and because they know that they will have trouble transferring employees out of their unit unless they give them high ratings. Soon everyone's rating is inflated to whatever the top one is. Except for a few souls who are out of favor and being punished for one reason or another.

It seems to me that ratings, if used at all, should be a private matter between a supervisor and a subordinate. The same function would be performed, without the stigma and corruption. Also, it would make discussions of performance much more relaxed and effective by removing the fear on one side and the anxiety on the other. The purpose of a discussion of performance is to provide feedback to an employee. Therefore, all that is really necessary is for the discussion to take place. Filling out forms and setting ratings that usually accompany evaluations add nothing to the function. For making sure that periodic discussions do in fact occur, all that is necessary is a simple statement to that effect, signed by the supervisor and the employee. All it needs to say is that a meeting took place at which performance was discussed.

Confusion is commonly generated during discussion of performance by mixing objectives and skills. In development work skills are secondary to objectives. Highly skilled people who consistently miss deadlines are not highly valued. However, for science and invention, objectives must be subordinated to skills. Here well-organized people who consistently meet objectives, but who do not enter novel, unknown futures are not highly valued. Managers who understand the difference and use appropriate policies can boost inventivity—and conversely.[3]

Another man's words say it clearly in an alternative way. He is Peter Carruthers, a former Head of the Theoretical Division at the Los Alamos National Laboratory.[4]

> *Ever since I've been here (LANL) there's been an increasing trend, both externally and internally, towards the illusion that you can "manage" science, whereas all you can really do is to get good people who are interested in the subject you want to develop. This increasing accountability at all levels of the federal establishment exudes a cold air that drives out the kind of neurotic and creative people that you need to make a breakthrough. There has to be a feeling of freedom and reward. You can't get good science out of people who recognize that they are being managed.*

NOTES

1. A. Szent-Gyorgyi, in *Perspectives in Biology and Medicine*, Vol. 15, Chicago: University of Chicago Press (1971) p.1.
2. Godfrey Hodgson, *The Colonel: The Life and Wars of Henry Stimson, 1867–1950*. New York: Knopf (Random House) (1990).
3. H. Levinson, "Management by Whose Objectives?," *Harvard Business Review* (July–August 1970).
4. *Science* (October 8, 1982) p.135.

13

Operational Needs of Research Organizations

Committees are, by nature, timid. They are based on the premise of safety in numbers; content to survive inconspicuously, rather than take risks and move independently ahead. Without independence, without the freedom for new ideas to be tried, to fail, and to ultimately succeed, the world will not move ahead, but live in the fear of its own potential.

—Prof. Dr. Ing. F. Porsche, 1989.

Operations here means the functions that facilitate the day-to-day activities of a research organization. There is a temptation to call these the "support services," but this demeans them and minimizes their importance, whereas I wish to emphasize their importance. As stated previously, research is a sequential process. Often the next step in the sequence cannot be started until the current step has been completed. Therefore, anything that can hasten the completion of the current step is an integral part of the work, not something that is in "support" of it.

One of the most foolish practices found in research organizations is cutting back on operations when budgets are tight. This slows the work of everyone; the high output researchers as well as the low. In fact, it slows the high output ones more than the low output ones. In large organizations it leads to overhead ratios that are excessive.

Operations include office work, shops, personnel, buildings and grounds, eateries, utilities, safety, chemical and physical analysis, purchasing, receiving and inventory, computations, communications, libraries, finances, transportation, and more. Some of these are more crucial than others, but most can lead to time delays so it is important that they all be done efficiently and with minimal intermediate steps.

Essential in getting research work done promptly are the shops that build various kinds of equipment; the purchasing operation that buys instruments, tools, and supplies; and the computing operation. Since they are essential to maximizing the rate of research and development output, these and perhaps other operations should lie as high as possible in the structure of the organization. Under most circumstances they should report directly to the director. They should certainly not be buried somewhere in the depths of the organization.

The people who work in the operational activities should be recognized as being essential to the success of the organization. Not as "hired hands." Their opinions should be solicited because they often have specialized information that others do not have. They should be included in all general meetings of the staff, in parties, and at other occasions. If they feel positively toward the organization they can often make suggestions that will improve either their own operation, or some other operation.

Since the workers who carry out the operations see the researchers actually at work, and the management does not, operations workers often have valuable opinions about personnel matters. They know who is skilled at doing experiments and who is clumsy. Who gets results with minimum elaboration and maximum finesse. Who is more interested in designing and building complex machines than in using them. And other information that is difficult to obtain otherwise.

No detailed discussion of research operations will be given here. There are books on the subject to which the interested reader is referred.[1] A model will described the great importance of efficient, and steady, operations in order to maximize the rate of output from a research organization—that is, its inventivity.

MODEL OF OPERATIONAL EFFICIENCY

A common complaint of researchers is that ups and downs in budgets have debilitating effects on the progress of their work. If this complaint were not so widespread it might be dismissed as self-serving comment. However, there is an underlying mechanism that accounts for the universality of the complaint, and indicates that more than unfounded opinion is involved.

Another commonly heard statement is that interruptions associated with meetings of committees, administrative chores, and the like, "leave no time to get research done." Although this response is partly subjective, it appears to be stronger than would be expected if the effect were proportional to the relative amounts of time that are involved. That is, the effect appears to be nonlinearly related to the cause. It is believed that there is a mechanism underlying this complaint that is closely related to the one of the previous paragraph.

The model that will be described is based on the simple, but not intuitive, fact that time lost during a slack period cannot be regained simply by increasing the effort at a later time by an amount equal to the amount of decrease in effort that occurred during the slack period. As a result, short- and long-term fluctuations in research efforts have markedly negative effects on average productivity. These effects are permanent unless the slack periods are deliberately counteracted by increased levels of effort that are larger than an amount equal to the foregoing decrease. Conversely, productivity should increase significantly if means can be found to reduce fluctuations in efforts.

Unlike manufacturing and some other kinds of work, research work does not lend itself to being speeded up by doing several operations in parallel. It involves overcoming a series of obstacles that must be approached in sequence—one after the other. This cannot be avoided because each subsequent obstacle does not become clearly defined until the preceding one has been overcome. Thus it is inherently sequential. Therefore, the rate at which research work achieves progress depends on the instantaneous effort that is put into it.

I shall begin by defining "effort" expended on research, E, as the difference between the total man-months available, M, and the nondiscretionary man-months, M^*, used for doing administrative tasks. These are related by:

$$E = M - M^* \qquad \text{with } M^* < M$$

so the effort becomes smaller as the nondiscretionary time increases.

As effort is expended, progress toward a goal is expected. The goal may be a completed study, or the completion of an experiment, the definition of an invention, the development of a prototype, or some other item. Progress will be represented by p; and the rate of progress, dp/dt by P. This rate will depend on the effort, E in a way that will be discussed shortly. Before discussing that, one needs to have in mind the effect of fluctuations of the effort.

The general effect is related to the "rowboat puzzle." Suppose there is a gently flowing river, and two rowboats manned by rowers of equal strength. Both start out at the same point along the shore of the river. One rows directly across the river and back to the starting point, or a distance $2w$, where w is the width of the river. The other rows directly down the river a distance w, and then back to the starting point. Do they both get back to the starting point at the same time? Or, does one get back before the other? It might seem that the one rowing parallel to the river would gain as much time when moving with the flow as he loses when moving against the flow, so both would get back at the same time. But this is not the case, as a little calculation will show. Now let us see how this simple principle applies for fluctuating effort.

When effort is expended causing progress to occur, a result, R will be reached after some time has passed. Two cases will be compared. The first is when the effort is steady at some base level, E_o. In the second case the effort fluctuates. To make the calculation as simple as possible, the fluctuation consists of a decrease in effort by an amount, e, for half the amount of time that it takes to reach R; and an increase by the same amount for the other half of the time. Thus the base level effort has a simple "square wave" superimposed on it. Other, more complex, patterns could be used but they would complicate the calculations without bringing out any new points. Next, consider how the rate of progress is related to the effort.

The most simple assumption would be that the rate of progress is proportional to the effort (the linear case). This might be appropriate for manufacturing types of work, but it is not realistic for research work. A more realistic dependence is one that was proposed by Shockley;[2] and is known to many people as the Boltzmann function:

$$P = P_m \exp(-d/E)$$

where P_m is the maximum rate of progress, and d is a parameter that measures the difficulty of achieving the result, R. The form of this function is shown graphically in Figure 13.1. It suggests a threshold of effort below which progress is slow. This threshold is a about $E = d/2$.

In the presence of the fluctuations in effort described, and the progress rate/effort function of the last paragraph, average rates of progress can be calculated and compared with the rates when the effort is steady. Figure 13.2 shows a map of the results of such calculations for various levels of relative difficulty, $D = d/E_o$.

It may be seen that for easy tasks ($D = 1/10$) small fluctuations have little effect on the rate of progress compared with the rate for steady effort. The

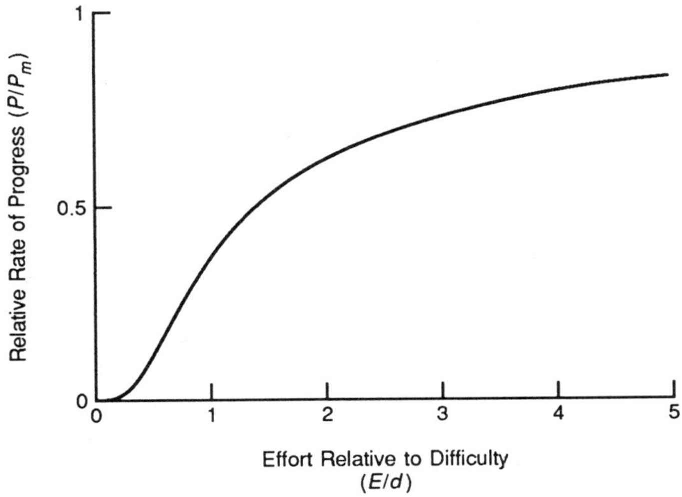

FIGURE 13.1. Form of the relationship between the rate of progress and the effort parameter. The latter is the effort, E, relative to the difficulty, d, of achieving a particular result. The inflection point where the rate of progress increases most rapidly with an increment of effort is at E/d = 0.5.

effect is large, however, for difficult tasks ($D = 10$). For example, for relative difficulty, $D = 2$, 50 percent fluctuations decrease the average progress rate by about 80 percent.

In this model the decreases in effort are symmetric with corresponding increases. More often, for real cases, one would expect asymmetry. That is, there would be downward fluctuations in levels of effort, but these would not be followed by corresponding increases. Thus the model treats the best case, showing that even then productivity suffers when the level of effort is not steady. And it shows that temporary slowdowns are not easily compensated; thus emphasizing the importance of efficient, reliable operations. Asymmetric fluctuations decrease the average rate of progress much more than symmetric ones.

Another feature of the model that is very important is the effect of nondiscretionary demands on rates of progress. The abscissa in Figure 13.2, the fractional size of fluctuations, has the level of effort in its denominator, as does the relative difficulty parameter, D. Therefore, increases in nondiscretionary demands tend to move a situation toward the right on the abscissa, and towards the left of the set of curves parameterized by D. Both of these tendencies decrease the average rate of progress.

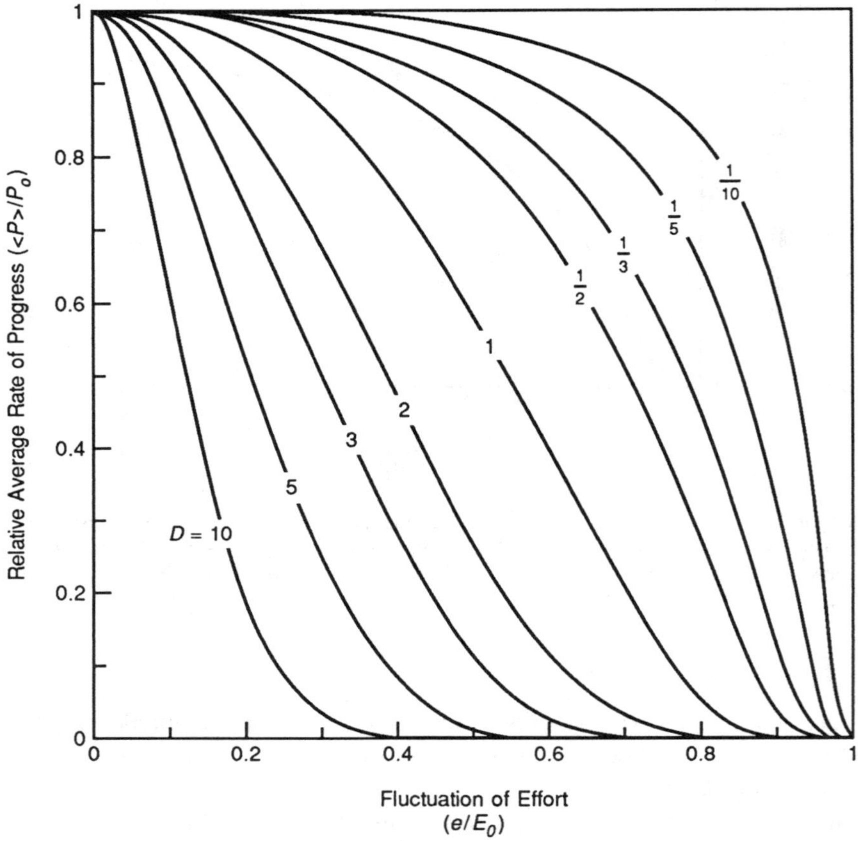

FIGURE 13.2. Calculated ratios of average rates of progress with fluctuations of effort to those without fluctuations. The amount of fluctuation increases toward the right. It is given as the ratio of the size of the fluctuation, e, to the size of a steady effort, E_o. The relative difficulty of obtaining a result, R, is represented by the parameter, D, which equals d/E_o where d is the difficulty parameter.

Effect of Organizational Size

The model just discussed suggests why large organizations tend to be inefficient in producing inventions compared with small ones. It is because interactions of various kinds increase with size. This decreases discretionary time, which reduces research effort, thereby reducing the rate of progress.

On the other hand, some organization is advantageous. Without it too much time is consumed with mundane tasks, and reinforcement of knowledge by interactions with other researchers does not occur efficiently. Thus there must exist an optimum organizational size. To clarify this, it can be described more explicitly, although the appropriate coefficients are difficult to evaluate.

To find an expression for the optimum size of a research organization, consider how the rate of progress for a particular researcher varies with the number of people, n in the organization. In order to enhance the rate of his or her work this researcher can seek the help of others. He or she has $n - 1$ choices so his or her rate of progress will tend to increase in proportion to $n - 1$. Assume that each researcher is independent of the others. Also, assume that n is large enough for the -1 to be neglected. Thus, for intermediate values of n, with a as the proportionality coefficient, and P_o as the base rate for the researcher working alone:

$$P = P_o(1 + an)$$

and the rate of progress increases with the size of the organization.

Objective evidence for this part of the size dependence is provided by the way in which technical papers are distributed when they have more than one author (Figure 13.3). The figure shows that the distribution peaks at two authors, indicating that productivity is increased by collaboration.

However, as n continues to increase, interactions will occur between the researcher and various other staff members that do not enhance the progress rate of his or her research work. Therefore, the fraction, f, of the total interactions that contribute to the rate of progress will tend to decline with increasing n when it becomes large. Changes in the fraction will be proportional to changes in the contributing interactions. That is, to $f(dn)$. Letting b be a coefficient, this means that:

$$df = -bf\ (dn)$$

and integration of this yields the declining fraction:

$$f = \exp(-bn)$$

so the expression for the rate of progress for the typical researcher becomes:

$$P = [P_o(1 + an)] \exp(-bn)$$

This indicates that the researcher's productivity first rises with n, and then

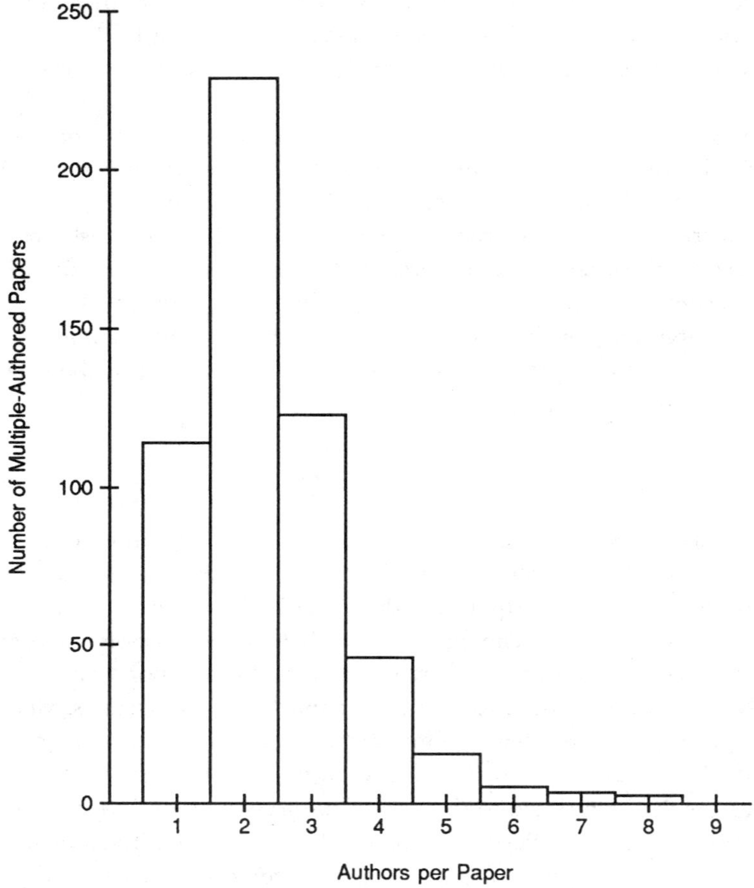

FIGURE 13.3. Distribution of papers with multiple authorship. Source: D. J. de Solla Price, Little Science, Big Science...and Beyond (New York: Columbia University Press 1986) p. 122.

declines toward zero as n becomes large. A schematic of this is shown in Figure 13.4. The maximum occurs when $n = (a - b)/ab$, which can be shown by setting the derivative of the previous expression equal to zero.

Unfortunately, values for a and b are not available so this is a qualitative model. Furthermore, a and b are not universal coefficients; they vary from one situation to another so the optimum size varies depending on the situation. For example, it seems apparent that objectives of small scale such as new optical components will have smaller optimum numbers of workers than large-scale objectives such as designing a hypersonic aircraft.

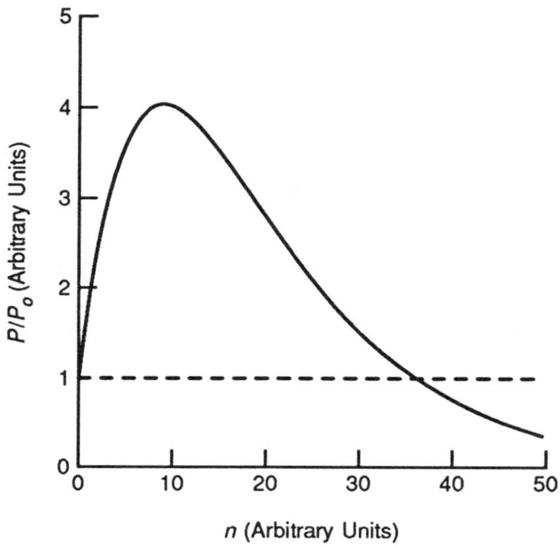

FIGURE 13.4. Schematic effect of organizational size on individual rate of progress. A plot of $P/P_o = (1 + an)\exp(-bn)$ with $a = 10b$.

There is evidence that the a coefficient is relatively large compared with b, so the optimum value of n is approximately equal to $1/b$. If some means can be devised to determine the optimum size, or to measure the b coefficient, the rest of the behavior can be estimated because the first part of the curve resembles the first quarter of the sine function.

Dropping the assumption that $a>>b$, if it is assumed that $P\ (n)$ is approximately a sine function up to the maximum point, then since the initial slope is a, it can be readily shown that the optimum value of n is $\pi/2a$. This gives another way to estimate the optimum size.

Another factor of interest is the amount of enhancement that might be expected from the size effect. That is, what is the maximum value of P/P_o? By substitution of the expression for the size, n at which the maximum occurs into the expression for P, it is:

$$[P/P_o]_{max} = (a/b)\ \exp\ [(b/a)-1]$$

Thus if $a<<$b this is zero, if $a = b$ this is one and there is no size effect, but if $a>>$b the ratio, $[P/P_o]_{max}$ becomes equal to a / b and therefore quite large. Thus the model is consistent with the idea that the non-contributory interactions described by the *b* coefficient, which reduce the rate of progress, must be minimized for organizations of any size.

Furthermore, since $[P/P_o]_{max}$ is independent of *n*, it cannot be increased by increasing the size. What is important is the ratio, *a*/*b*— not the size itself. But it must be kept in mind that the size must properly match the task at hand. This is the most important conclusion the can be reached from this model: that size alone does not enhance the rate of progress. The amount of enhancement that occurs depends on the kinds of interaction that go on inside the organization. The *a*-type need to be maximized, while the *b*-type are minimized. The general ways for doing this are to make the researchers as independent operationally as possible, and free to interact at random without inhibiting territorial boundaries. This is not a new recommendation, but it is made explicit by this model.

NOTES

1. V. P. White, *Handbook of Research Laboratory Management* (Philadelphia: ISI Press, 1988).
2. W. Shockley, "On the Statistics of Individual Variations of Productivity in Research Laboratories," *Proc.IRE*, Vol. 45 (1957) p. 279.

14

BENEFITS FLOWING FROM RESEARCH ORGANIZATIONS

The presence of a body of well-instructed men, who have not to labor for their daily bread, is important to a degree which cannot be overestimated; as all high intellectual work is carried on by them, and on such work material progress of all kinds mainly depends, not to mention other and higher advantages.

—Charles Robert Darwin *The Descent of Man* (1871).

In Chapter 2 the overall value of research and development was discussed. But why not buy the results of research activities instead of underwriting the expenses of supporting one's own organization? This question and related ones are approached here by factoring the overall activities of an organization into its component parts, and then presenting proposals for estimating their values in monetary terms. This will give some sense of how the various benefits are proportioned relative to the whole. Each part will have a value that is highly dependent on both its level of quality, and on the quality of the whole.

Since many factors are involved, the first step is to identify as many of the component factors as possible. This will be followed by estimating the values contributed by each of them. These estimates will be only approximate, but they may improve the clarity of thinking as it is devoted to the values of doing research. Also, attempts to improve research effectiveness can be better organized if it becomes possible to measure, or estimate, the values of the component factors. In order to approach the problem at all it is necessary to treat the component factors as if they were independent. In reality, it is their interactions that make a research organization lively, productive, and effective. Thus their interdependencies strengthen the values of the individual factors in successful organizations, whereas they tend to decrease the individual values in less successful ones.

We shall start by discussing the functional factors that comprise the research enterprise, and then consider how their values can be monetarily quantified (in terms of dollars). This process will help to identify some of the parameters that affect the values, and therefore suggest means for improving them.

One way of splitting the overall activity into its individual functions is illustrated in Table 14.1. The column on the left in this figure consists of the "line" or primary functions, while the one on the right lists the "staff" or secondary functions.

EVALUATION OF PRIMARY FUNCTIONS

The very essence of research is *learning*. People doing research read, study, analyze, discover, measure, and synthesize all kinds of technical information. They collect this information from the technical world at large in a variety of forms. The skill with which they collect it, digest it, and disseminate their conclusions can be judged in various qualitative and quantitative ways (Table 14.2).

As was discussed in Chapter 4, a number of sequential steps must each be completed before the overall learning function can be considered to have been done successfully. Therefore, those who are best at performing the learning function are very much better than those who are less good.

If any of the essential steps is not completed the function is not performed. Some of the steps that are involved are as follows:

TABLE 14.1. Functions of a research organization.

Line Functions	Staff Functions
Learning	Recruiting
Inventing	Advertising
Developing	Consulting
Piloting	Educating
Licensing	Forecasting

TABLE 14.2. Methods for judging learning outputs.

Qualitative Measures
Manuscript Acceptances
Lecture Invitations
Literature Citations
Esteem of Colleagues
Honors Received
Quantitative Measure
Societal Support of Scholars

1. Select a specific topic for study,
2. Work at the study with vigor and thoroughness,
3. Organize and/or synthesize the acquired knowledge,
4. Decide when to stop studying and begin writing down the results,
5. Write with adequate speed and clarity,
6. Submit written manuscript for review by peers and/or supervisors,
7. Persevere in spite of criticisms by peers and/or supervisors, and
8. Make corrections and revisions

Because of the large spread in the effectiveness of individuals in executing these steps, it is not possible to assign a quantitative value to the best efforts. However, the value of an average effort can be estimated. In its purest form, the learning function is done by scholars in universities. At equilibrium, society pays for supporting its scholars at a rate that approximates their dollar value to society. Currently, this amount (salary plus overhead) averages about $200,000 per man-year for high-level technical scholars. A lower limit can be placed on the value of a man-year of learning by equating it with the cost charged to a student, which currently is about $25,000 per year in a first-class university. The root-mean-square average of these two limits is $143,000 per man-year.

Concepts and measured data that are obtained through learning are combined to form new ideas by talented people doing *inventing*. The inventions may take the form of new materials, processes, devices, systems, or software. Once again the most talented people produce far more inventions than those less talented. The steps that must be accomplished in sequence to produce an invention are even more demanding than those needed to produce a scholarly work. For a typical laboratory, approximately 5 percent of the staff produces 10 times as many inventions as 50 percent of the remainder (see Figure 4.2). Measures of invention productivity are listed in Table 14.3.

TABLE 14.3. Methods for judging invention productivity.

Qualitative Measures
Invention Memoranda
Issued patents
Patents practiced
Quantitative Measure
Value of average patent

Inventions are converted into concrete pieces of property by obtaining patents that describe and protect them. The process begins when the inventor(s) writes down an informal description of the invention; sometimes called a memorandum of invention, or an invention disclosure. This is usually based on information recorded in a notebook. These are placed into files. Most of them are not pursued for one or more reasons:

1. Further thought indicates that the idea will not function properly.
2. It is realized that the idea lacks novelty.
3. The inventor wishes to work on ideas of higher priority.
4. Reduction to practice would be difficult.
5. The idea does not mesh with the priorities of the management.

If an invention memorandum or disclosure is pursued and it continues to show promise after initial experimental work, and a preliminary study of the prior work has been completed, it forms the basis of a patent application to be filed at the U.S. Patent Office. After the passage of one or two years, a patent is issued and is good for seventeen years.

Issued patents may be practiced by a commercial enterprise, or may be licensed to other companies, or they may lie dormant. It seems reasonable to suppose that the value of an average patent is related to the cost of obtaining it. Otherwise experience would lead to the conclusion that the cost exceeds the value and there would be no filings. The current cost is estimated to be about \$25,000. In order to make up for the patents that turn out to have little or no value, an average one needs to have much greater value than its cost. As a consequence of the rule-of-tens, this amounts to \$250,000. About 1 percent of all of them are far more valuable than the average—that is, about \$2,500,000. Note that this value relates only to the patent itself. It may be the first step in an enterprise whose ultimate value is many orders of magnitude greater than this amount.

After an invention has been conceived, considerable *development* must be done in order to match it to the needs and demands of potential users. The direct costs of such work should be viewed as investments made to generate future returns. Thus the value of such work can be estimated by calculating the present values of estimated cash flows. In order to identify the important parameters that are associated with the costs and values, consider a simplified model. It should be understood that this model is too simplified to represent any real situation accurately. It is only intended to be schematic.

The model is illustrated by Figure 14.1. In this schematic figure, a development effort, D_o is invested at time period one. If it were not used for the development work this investment could be invested elsewhere and earn interest at a rate, i (the real rate—not the nominal rate) so that after n time periods had passed it would be worth (with continuous compounding):

$$D_n = D_o \exp(in)$$

D_o consists of both capital expenditures and operating expenditures. To develop a business of a particular size in terms of sales requires a capital investment roughly equal to the expected sales, and operating expenses per year roughly equal to half of the expected sales per year.

The development work is expected to create a business that generates earnings beginning at time period, n_s, amounting to E_o, and growing at a rate, g. Thus, the earnings at the nth time period are:

$$E_n = E_o \exp[g(n - n_s)]$$

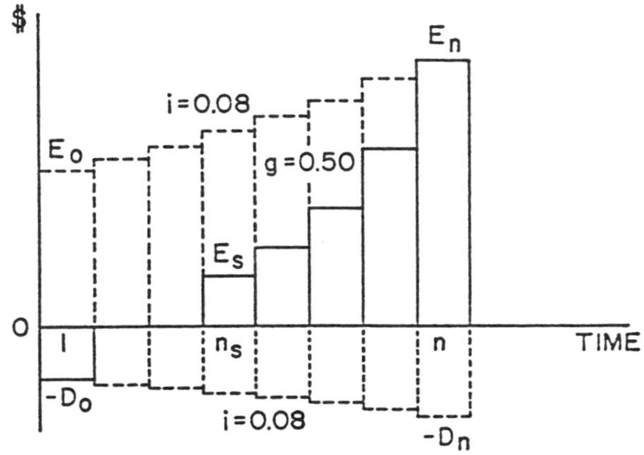

FIGURE 14.1. Schematic representation of development model.

These earnings can be reinvested during the period n_s to n. However, in order to keep the model simple, discount these earnings back to the initial period to get their present value.

Earnings obtained at some time in the future are worth less than the same amount of money held at present because they could be invested for interest yield if they were held at present. Thus the present value equals the earnings divided by the discount factor, $(1 + i)^n$, and the total present value of the earnings is the sum of such terms for the periods when $n > n_s$.

The net present value (NPV) is the development cost (negative present value) plus the discounted earnings:

$$NPV = -D_o + \sum_{n=n_s}^{m} [E_n/(1+i)_n]$$

not including growth. This expression illustrates what happens to the net present value when growth is not included; namely, as the number of time periods increases, the NPV approaches a constant value because each successive term in the sum is smaller than the previous one. This indicates the disadvantage of development projects that have only a one-time effect—for example, a one-time cost reduction.

Pressure is often exerted on research organizations to do work directed at cost reductions. Such efforts are not good investments, however, unless they lead to growth. This is not to say that they are necessarily bad investments; only that they are distinctly less good than developments that have growth prospects.

When growth is included, this expression becomes:

$$NPV(g) = -D_o + \sum_{n=n_s}^{m} \{E \exp[(g-1)n - gn_s]\}$$

Now the successive terms in the sum increase if the growth rate, g, is greater than the interest rate, where both rates are real rather than nominal. Thus the NPV increases indefinitely in this case.

Another thing that may be noted about this expression is that for a given growth rate, the time at which earnings begin, n_s, plays a very important role. In order to make the net present value as large as possible, it is essential for n_s to be as small as possible.

A lower limit (approximate) can be placed on the value of development work by assuming that on average it leads to moderate success, meaning that it returns its direct costs plus the interest lost by not investing the direct costs elsewhere—say 10 percent.

The step that follows the development of a new material, process, device, etcetera is *piloting*. This consists of moving it out of the relatively safe harbor of the research environment into the harsh sea of the marketplace. One aspect of this new environment is that the value of the activity becomes readily measurable by means of a profit and loss statement.

Another way of looking at the value of piloting is that it is essentially a service, so its value equals its cost because it would be necessary to purchase the service externally if it were not done internally.

Income derived from inventions and their associated patents may come from *licensing*. The value of this is whatever its direct income is minus the costs of collecting the income.

EVALUATION OF SECONDARY FUNCTIONS

An important result of the presence of an effective research group within its structure is that a corporation, or other organization, becomes able to attract and *recruit* incrementally superior individuals in technically related areas. The research group can help to identify such people, to attract them to interviews, to judge their qualifications, to persuade them to become employment candidates, and to keep them from moving elsewhere after they have been hired.

Although most organizations assert that their employees are above average, this cannot be true in reality. Some must be below average, if others are to be above average. In reality, most are close to average. Organizations that exert a well-executed effort may be able to rise above average. One effective way of doing this is to recruit employees initially into a research organization and redistribute at a later time. This method has served several large organizations such as AT&T, DuPont, and General Electric very well.

The value of the incremental improvement in quality of personnel induced by the presence of a research department can be estimated by taking it to be (on average) the difference in salary between the lowest and highest performance quartiles for technical personnel. This amounts to about $50,000 per man-year, keeping in mind that this seriously underestimates the real incremental value of a truly superior recruit.

Suppose an organization employs 1,000 technical people whose recruitment has been influenced by its research group. If the influence raises the annualized value of the average employee by 10 percent, or about $15,000, then the research group deserves a credit of about $15,000,000 per year; an amount that probably covers the research group's costs without looking further for benefits!

Closely related to the recruiting benefit is that of *personnel development* for technical management. This value is the incremental cost of external versus internal training, plus the employment expense that would be associated with recruiting external employees.

Opportunities for substantial institutional *advertising* are often provided by research organizations. The general public is interested in new ideas, discoveries, and inventions. It will give generous attention to these while ignoring announcements of business activities. Novel research results often lead to interesting photographs as well as news texts. Also, the publishing activities that are an important routine of research work provide further institutional advertising. Lectures and other kinds of presentations also have advertising value.

The one-time page rate for advertising in a professional or trade journal averages about $5,000. The actual rates depend on the amount of circulation the journal receives. Under the reasonable assumption that this approximates the value of the advertising (on average), it provides a means for quantifying the values of technical photographs, scientific papers, news items, and the like.

For specialized topics, a research staff can often provide internal *consulting* services to other technical areas of an organization. These services may range from making measurements, or analyses, using specialized equipment, to suggesting sources of information, to providing computations related to particular designs and operations. The present cost of purchasing such work externally is about $800 per day.

A continual activity within any large organization is *educating* its staff members. This education may pertain to new techniques or to an increase in depth of knowledge that suddenly acquire new importance, or it may be associated with attaining of greater breadth of knowledge by various individuals. All of these events increase an employee's value and provide desirable background for future leadership. The value of this activity is approximately equal to the incremental cost of obtaining it by hiring pretrained persons from the outside (approximately 25 percent of base salary per person), plus the cost of the incubation period needed for the outsider to become acquainted with the new environment (say 75 percent of base salary per person).

Numerous situations arise in which *forecasting* needs to be done. Much of this has a substantial technical content, which needs background knowledge, analytical techniques for extrapolation, plus judgment based on experience. If it is done competently, it can help an organization avoid bad situations of the future, or to move toward good ones. To purchase this service costs about $50,000 for an average case.

Although the values of the various secondary functions of a research organizations do not cover its total costs, they do offset a significant fraction of them.

BALANCE OF COSTS AND BENEFITS

In order to see how these factors compare with one another, Table 14.4 lists them for a hypothetical organization of modest size. It employs 75 professional scientists. Each of these is supported directly by one person and indirectly by another for a total of 225 people.

Much overlapping of functional work occurs within research organizations, so the functional descriptions of people cannot be sharply categorized. It is assumed that managers have average effectiveness so their values are in equilibrium with their costs. The same is assumed for the people who do development and piloting work because the value of their work depends on future business events. The current cost of a research worker (includes support personnel, rent, equipment, and overhead) is assumed to be about $200,000. This will vary with time, and overhead rates.

In my experience, an organization of the size cited can be expected to produce the following: about 25 patentable inventions per year; successfully recruit of about 10 professionals of high quality; 75 technical publications; 20 man-days of internal consulting; educational programs for 5 people who will make transitions to other than their present jobs; and involvement with about 10 forecasting cases.

Table 14.4 shows that the costs and benefits are approximately equal. No special significance is attributed to this because other assumptions could produce a different result. It does indicate, however, that there is no large imbalance. In other words, it is believed that these estimates verify that an average research organization pays its own way. Since the cost of a better than average organization differs little from that of an average one, the better one does far more than pay its way.

TABLE 14.4. Cost/Benifit balance sheet (see text for assumtions; dollar amounts in thousands).

BENEFICIAL VALUES		
Primary functions:		
Learning	(20 persons @ $143)	$2,860
Inventing	(25 patents @ $100)	2,500
Developing	(35 persons @ $200)	7,000
Piloting	(5 persons @ $200)	1,000
Licensing	(2 licenses @ $100)	200
Managing	(10 persons @ 200)	2,000
		$15,560
Secondary functions:		
Recruiting	(10 recruits @ $50)	$ 500
Advertising	(75 publications @ $5)	375
Consulting	(24 days @ $0.8)	19
Educating	(5 persons @40)	200
Forecasting	(10 cases @ 50)	500
		1594
	TOTAL	$17,154
COSTS	(75 persons @ 200)	$15,000

These estimates indicate that the staff functions have significant value—about 20 percent of the total.

The importance of doing development work effectively and efficiently is clear since it has the greatest single value of the functions listed (or cost, if it is not done well).

15

So What?

If arithmetic, mensuration, and weighing be taken away from any art, that which remains will not be much.

—Plato.

When I was doing laboratory research, a visitor came by one day; a candidate for employment who was being shown around and interviewed. I proudly explained some of my hard won results to him. When I had finished, he asked, "so what?" I was nonplussed by this aggressive question, and sputtered out a muddled and inadequate answer. Needless to say, he did not become highly placed on my list of employable candidates. He has gone on to do distinguished, but to me esoteric research work. So we're even.

Although it is a tough question, "so what?" is a legitimate one. After fourteen chapters, what can be concluded? A common statement is that research work is unmanageable. This is not a conclusion of this book. However, it is concluded that several of the directions in which the practice of research management is headed are badly misguided. In conversation one might say stupid. Why? Because these directions violate simple, albeit sometimes subtle, principles. These principles, where they can be found, are more than opinions, because they are founded on observable facts. I hope that this is the prime message of the book; sociological principles can be found if a search is made for them. They are not easy to find, and the ones outlined here are rudimentary, but they exist.

In this age of organizational gigantism, expositing a position that emphasizes individuality may seem quaint, but it is justified by observations of the

Pareto distribution law in action. Individuals populate the high end of the distribution—not teams. These individuals are the ones who change society through their research discoveries. Finding means to help them proceed into and up through the Pareto distribution is the primary job of management. Unfortunately, it is a job that rarely gets executed.

An attempt has also been made to emphasize that it is not the total number of inventions made per unit of time that is important. It is the inventions divided by the cost: the inventivity. When this becomes too low, companies and other organizations lose their competitiveness. They cannot afford to spend the surplus funds that they may have on an effort to make inventions that do not yield an adequate return on the investment. If the situation is not changed effectively, such companies are likely to practice obsolete technologies and eventually become bankrupt. Thus, inventivity is a critical parameter for companies, governments, and societies.

The only truly effective means that management has for increasing inventivity is to hire, or put in place, the right people. Other means can "trim" it, but cannot have a primary effect. This is because it is useless to ask people who are uneducated, or inexperienced, to practice a technology at a high enough level of sophistication to produce inventions. Therefore, the pool of people from which choices can be made will contain mostly people who do not have the characteristics of inventors. It will also contain a dilute concentration of people who do have the characteristics. Picking them out should be the primary activity of the management. Not picking out projects; picking out people.

Paraphrasing a statement by Lukes,[1] the hard fact is that the overwhelming majority of technologists are not capable of making inventions to any significant extent; just as there are many well-trained musical arrangers but there are very few really good composers (or improvisors).

Inventors can best be identified by their own statements about their previous works. A person who talks about his or her inventive accomplishments, or how much he or she enjoys the invention process, and nothing else, is unstoppable. If this is combined with solid knowledge of a technology, and some tools are made available to the person, the odds are high that inventions will ensue. On the other hand, if the person talks about other things, such as sports, family, community, or others, no amount of cajoling or encouragement will lead to significant inventions.

Although there are few ways to increase inventivity, perhaps only the one previously outlined, there are many ways to decrease it. Second-guessing the inventor is probably most effective. Inventing is a discouraging activity. Its path is strewn with failures. Successes are rare. Inventors feel that nature (the second-guesser *par excellence*) mistreats them in abundance. They feel no need for more mistreatment from management. The conjugate of this is that inventors welcome sympathetic attention, constructive comments, visits to their laboratory or office, and other indications that management appreciates their efforts. Legend has it that Willis Whitney visited everyone in the General Electric Laboratory daily even after it had become sizeable. This was during its period of highest inventivity. Was there a correlation? Probably.

The methods that can used to improve inventivity involve helping the inventors get organized. Not overtly, but by indirection. By expressing curiosity about what they are doing, why they are doing it, and what their timetables are. By asking them to write brief reports and present brief accounts of their past doings. By improving the morale of people around them. By making it as simple as possible for inventors to record ideas that may lead to patentable inventions. Such ideas, even the most brilliant, start as delicate embryos and should be incubated, not assailed by an insensitive bureaucracy.

As in most other endeavors, time is the enemy; especially as it is expressed through the time value of money. High real interest rates drive everyone to shorten their time horizons. There is no choice. It is not a matter of taste. This should be clear to all in a market-based economy. It is hidden in centrally planned economies, but eventually emerges and has its destructive day. Cheap money is the friend of the inventor, or the developer, of new technology.

The amount of time that it takes to get from the conception of something new to its acceptance in the marketplace is surprisingly long. Intuition suggests that good ideas are put to work quickly, but this is not true in real cases. For the 60 cases listed in Appendix III, the average time lapsed was 22.5 years, and the mode was about 12 years. But after 12 years, if the real interest rate is 5 percent, a unit of money is only worth 0.56 units. That's the downside. The upside is that if "normal" net earnings at a rate of 15 percent start at the third year on one unit of sales, and the sales growth rate is 20 percent, the cumulative earnings (taking credit for reinvestment of earnings) is 4.57 units. This may be compared with 1.80: the value of one unit after 12 years at the given interest rate.

There is some danger that the whole research enterprise will founder if its relative costs are not controlled, but if this does not occur the outlook for further progress is brilliant. Photonics is still in its infancy, and it will cause more profound changes in the world than electronics has and continues to do. Complex integrated systems represented by robotics will have profound effects. It is not possible to even imagine the transformations that will come from molecular biology. It can also be expected that what is sometimes called quantitative sociology will change the way the world operates. These fields and many others will provide bountiful opportunities for research and inventors. The higher the inventivity becomes, the more rapid progress will be.

NOTES

1. Robert M. Lukes, "Masquerade of Undirected Research," *Science*, Vol. 159 (1968) p. 34.

Appendix

I

OPTIMUM RESEARCH SPENDING MODEL

We start with the expression for the correlation between the P/E ratio and relative research spending (R/S):

$$P/E = (P/E)_o + K(R/S) \tag{1}$$

where P = price per share, E = earnings per share per year, K = slope of correlation line, R = R&D spending per share per year, S = sales per share per year, and $(P/E)_o$ = intercept of correlation line at $(R/S) = 0$. It should be understood that this expression is only valid for $(R << E)$ because in addition to R there are other demands on earnings.

Next consider two earnings ratios. $(E/S)_b$ = earnings prior to research expenditures; and $(E/S)_a$ = net earnings after taxes. Let T be the tax rate expressed as a fraction. Then a tax factor, t can be defined as:

$$t = 1 - T \tag{2}$$

This can be used to relate the earnings ratios and the relative research spending:

$$(E/S)_a = [(E/S)_b - (R/S)]t \tag{3}$$

The relative market value of a company is (P/S). This may be written:

$$(P/S) = (P/E)(E/S)_a \tag{4}$$

or Equation (1) times Equation (3):

$$(P/S) = t\{(P/E)_o(E/S)_b + [K(E/S)_b - (P/S)_o](R/S) - K(R/S)^2\} \quad (5)$$

Thus the market value is at first enhanced by increasing (R/S), and is then decreased when the third term in Equation (5) begins to dominate. The maximum market value, and the optimum research spending, are readily found by differentiating Equation (5) and setting the derivative equal to zero. The optimum research spending is:

$$(R/S)^{opt} = \{(E/S)_b - [(P/S)_o/K]\}/2 \quad (6)$$

To obtain the maximum market value this is substituted into Equation (5), yielding:

$$(P/S)^{max} = t\{(P/S)_o(E/S)_b + K/4[(E/S)_b - (P/S)_o/K]^2\} \quad (7)$$

A p p e n d i x

II

THE MARKET SHARE VERSUS RANK RULE

The distribution rule is as follows. Let M_j be a fractional share in a market with n participating companies. The rank index for a particular company is j. For the company with the largest market share, $j = 1$, and for the one with the smallest share, $j = n$. Thus j = 1,2,3,...., n. The function that yields the market shares is:

$$M_j = (1/n) \sum_{i=j}^{n} (1/i)$$

For example, consider a product produced by three companies ($n = 3$). Then the three shares that correspond to them will be:

$$M_1 = (1/3) \sum_{1}^{3} (1/i) = 11/18 = 61\%$$

$$M_2 = 5/18 = 28\%$$

$$M_3 = 1/9 = 11\%$$

Figure 3.3 is a table of values calculated from this distribution rule for various sets of companies up to a set of 10. Figure AII.1 is a graphic representation of the table.

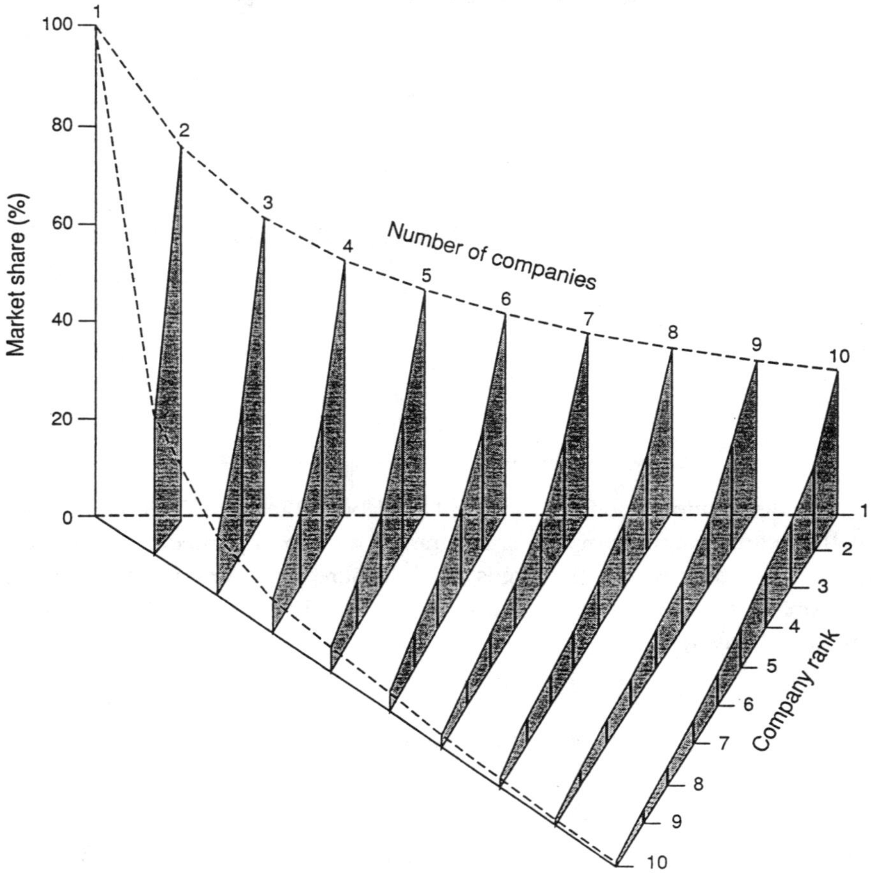

FIGURE AII.1. Graph of market share rule for sets of companies (up tp ten). Each set shares a particular market. The upper dotted line indicates the share (plotted vertically) of the market leader in each set. The lower dotted line indicates the share of the market laggard in each set.

Appendix

III

MARKET PENETRATION TIMES

Substitution	Mid-Year of Transition	Length of Transition (Years)	Ref Number
Open-hearth/Bessemer steel	1907	38	1
Organic/inorganic insecticides	1946	17	1
Electric-arc/open-hearth steels	1947	43	1
Synthetic/natural tire cord	1948	16	1
Titania/lead-zinc pigments	1949	24	1
Detergent/soap (U.S.)	1951	8	1
Synthetic/natural rubber	1956	53	1
Synthetic/natural leather	1957	52	1
Margarine/butter	1957	51	1
Sulfate/tree-tapped turpentine	1959	38	1
Plastic/hardwood flooring	1966	23	1
Plastic/wood boat hulls	1966	18	1
Glass/wood boat hulls	1960	10	1
Water/oil-based paints	1967	39	1
Basic oxygen/open-hearth steels	1968	10	1
Synthetic/natural fibers	1969	53	1
Plastic/metal auto bodies	1981	15	1
Steam/sail-boat propulsion	1880	54	2
Oil/coal-boat fuel	1924	58	2
Power/manual auto steering	1962	20	2
Refrigeration/natural auto cooling	1968	15	2
Vacumn/mechanical rug cleaners	1952	29	2
Electric/manual washing machines	1943	40	2
Electric/manual dishwashers	1980	31	2
TV/radio entertainment	1953	16	2

Substitution	Mid-Year of Transition	Length of Transition (Years)	Ref Number
Color/black & white television (U.S.)	1972	8	2
Turbofan/turbojet aircraft engines	1961	5	2
Reforming/cracking catalysts	1955	5	2
Polyethylene/glass bottles	1963	18	3
PVC/glass bottles	1978	20	3
Diesel/steam locomotives	1949	9	4
Outboard/inboard boat engines	1962	8	4
By-product coke ovens	1908	34	5
Tinned containers	1936	4	5
Shuttle cars	1948	15	5
Car retarders	1940	34	5
Trackless mobile loaders	1942	19	5
Continuous mining machines	1951	12	5
Pallet-loading machine	1953	11	5
Continuous steel strip mill	1932	12	5
Centralized traffic control	1937	29	5
Continuous annealing	1948	27	5
High-speed bottle filling	1958	16	5
Radial/bias-ply auto tires	1976	5	6
Silicon/germanium semiconductors	1963	2	6
Integrated/discrete electronics		7	6
Electronic/mechanical cash registers	1974	4	6
Crushed rock/granite block ballast		10	7
Frozen/canned fruit juice	1950	30	8
Canned/bulk fruit juice	1941	20	8
Residential electrification	1928	45	9
Ultrasound imaging	1975	7	10
CT head imaging	1976	6	10
CT whole body imaging	1977	5	10
Mammography	1974	8	10
Surfaced/ "dirt" roads	1945	58	11
Metal/wood ships	1910	50	12
Autos/horses(nonfarm)	1916	14	12
Disc/drum auto brakes	1970	7	12
Unleaded/leaded gasoline	1980	15	12
	Ave	22.5	

NOTES

1. J. C. Fisher and R. H. Pry, "A Simple Substitution Model of Technological Change," *Tech. Forcast. Soc. Change*, Vol. 3 (1971) p. 75.
2. A. W. Blackman, "The Market Dynamics of Technological Substitutions," *Ibid.*, Vol.6 (1974) p. 41.
3. M. O. Stern, R. N. Ayres, and A. Shapanka, "A Model for Forecasting the Substitution of One Technology for Another," *Ibid.*, Vol. 7 (1975) p. 57.
4. N. N. Sharif and A. K. Haq, "Determinants for Forecasting Technological Substitution," *Ibid.*, Vol. 13 (1979) p. 59.
5. E. Mansfield, "Technical Change and the Rate of Initiation," *Econometrica*, Vol. 29 (1961) p. 74.
6. R. N. Foster, "Innovation," (New York: Summit Books, 1986) p. 141.
7. A. D. Moore, *Invention, Discovery, and Creativity* (New York: Doubleday and Company, 1969) p. 108.
8. S. O. Murray and J. H. Rankin, "Use Diffusion: An Extension and Critique," *Tech. Forcast. Soc. Change*, Vol. 16 (1980) p. 331.
9. J. M. Robinson, "Technological Learning, Technological Substitution, and Technological Change," *Ibid.*, Vol. 18 (1980) p. 39.
10. C. Easingwood, V. Mahajan, E. Miller, "A Nonsymmetric Responding Logistic Model for Forecasting Technological Substitution," *Ibid.*, Vol. 20 (1981) p. 199.
11. C. Marchetti, "On the Role of Science in the Postindustrial Society," *Ibid.*, Vol. 24 (1983) p. 197.
12. N. Nakicenovic, "The Automobile Road to Technological Change," *Ibid.*, Vol. 29 (1986) p. 309.

Appendix

IV

THE FORM OF MARKET PENETRATION CURVES

A remarkable fact about market penetration curves is their constancy. The shape is a simple *S*, and is always the same, although the time involved may be extended or contracted depending on various factors. Thus the process of market penetration appears to be a stochastic one. That is, it is made up of a sufficiently large number of small random events that the form of its path is statistically determined.

Market penetration is analogous to the diffusion of a concentration gradient—not to a propagation process. It does not spread along direct paths like acoustic or optical waves. Rather it moves in a complex series of nearly random steps. Some of the steps advance its cause, others retard it. The larger the economic incentive associated with it the faster it advances. By economic incentive is meant the *difference* between the utility of the newer product compared with that of the older one.

The diffusion occurs in both time and space. From the place where the technology is first developed, it diffuses to other places. When it reaches a new place, its use begins to grow there. The growth follows the simplest of kinetic rules: the rate of growth is proportional to the amount of use of the new technology that is already occurring (first-order kinetics). Thus the amount of use increases exponentially with time. Later the process saturates. Encounters between existing users and potential users become less frequent because most potential users have already been converted. Now the growth rate declines; approaching zero asymptotically.

Let the fraction of a market that is being penetrated be f. It varies between 0 and 1. The former of these values refers to the initial situation; the latter refers to the time when the market has been completely taken over by the new product. Then initially f increases a rate, df/dt, that is proportional to f. Let us call the proportionality constant, A. But the rate at which f increases is also affected by the declining amount of market that remains to be penetrated, or $1 - f$. Thus the overall process can be described by the following differential equation:

$$df/dt = Af\,(1 - f) \tag{1}$$

This most simple form can be elaborated in various ways, but it is remarkably effective as it stands.

Equation (1) cannot be integrated near $f = 0$, so two approaches have been used to set appropriate boundary conditions for it. one is to set f equal to a small value, f_o, at zero time. This yields the solution:

$$f\,(t) = A\{1+[\,(A/f_o) - 1\gamma e^{-At}\}^{-1} \tag{2}$$

which has the disadvantage that the actual time at which $f = f_o$ is difficult to determine.

The other approach due to Fisher and Pry (Reference 1, Chapter 9) is to assume that the penetration curve is symmetric so Equation (1) can be integrated from $f = 1/2$, where the time is t_m, to $f = 1$ at $t =$ infinity, and to $f = 0$ at $t = -$ infinity. This gives:

$$f\,(t) = 1/2[1+\tanh A(t-t_m)] \tag{3}$$

which also has a disadvantage in that no starting time is defined because f approaches zero asymptotically as t goes to minus infinity.

Because of the difficulties near $t = 0$, Equation (1) is most useful near $t = t_m$ where the rate of penetration reaches its maximum value:

$$(df/dt)_{max} = A/4 \tag{4}$$

In this region, the behavior can be further simplified by linearizing the curve as in Figure 8.1. It can then be characterized by two time parameters. One is t_m at which $f = 1/2$. This specifies the center of the historical time period of the penetration process. The other is the inverse of the slope at $t = t_m$; or $4/A$. This describes the time interval during which most of the process occurs.

INDEX